W9-CLL-302

MATRICES IN CONTROL THEORY

with applications to linear programming

MATRICES IN CONTROL THEORY

with applications to linear programming

S. BARNETT

Senior Lecturer in Engineering Mathematics,
University of Bradford

VAN NOSTRAND REINHOLD COMPANY
LONDON

NEW YORK CINCINNATI TORONTO MELBOURNE

VAN NOSTRAND REINHOLD COMPANY
Windsor House, 46 Victoria Street, London S.W.1

INTERNATIONAL OFFICES
New York Cincinnati Toronto Melbourne

Library of Congress Catalog Card No. 76-160195
ISBN 0 442 00581 4

First published 1971

Printed in Great Britain by
Butler and Tanner Ltd, Frome and London

He who does not increase his knowledge, decreases it. Say not, when I have leisure I will study; perchance thou wilt have no leisure.

HILLEL.

Preface

Matrices play a fundamental role in the so-called modern theory of linear control systems, and interest in this field has been growing considerably in recent years. In consequence a large number of new results has been published in journals and reports, and this often means that valuable time has to be spent searching through the literature for some required fact. The primary aim of this book is to try to reduce this rather wasteful expenditure of effort by presenting a collection of some of the more interesting and useful recent developments in the applications of matrices to control theory and related topics, especially linear programming. The emphasis is very strongly on results which have not previously appeared in book form, and I have done my best to present this varied material in a coherent and unified fashion, although each chapter is very largely self-contained. In order to make the results accessible to as wide an audience as possible, a special feature of this book is that no knowledge of the concepts of vector spaces is required, all development being entirely in matrix terms. In this sense the treatment throughout is elementary and essentially straightforward. I am aware that results of greater generality are sometimes excluded because of this restriction but, in Britain at least, few non-mathematicians seem happy with an abstract approach.

It is my main intention, then, that this book should prove useful as a convenient source of reference for engineers, scientists and applied mathematicians working in control theory and related areas involving applications of matrices. In accordance with this aim each chapter contains a large number of references, and these are supplemented by a short list of selected additional references at the end of the book. However, another noteworthy point is that I have given worked solutions to all the exercises set at the end of each section, in the hope that this will also make the book appeal to senior undergraduate and post-graduate students as a textbook for self-study and as a stimulus to research. For this reason many exercises are not merely routine but often contain further developments and amplifications. Throughout I have given proofs of theorems only where these can be presented in a relatively straightforward manner; otherwise on grounds of conciseness and clarity I have referred the reader to the original source.

I have supposed that the reader has a good general background knowledge of matrix theory and some acquaintance with the basics of linear control theory and linear programming. However, for easy reference I have provided short summaries of important points in the form of appendices at the end of the book, together with suggestions for introductory reading. Of course these notes are no substitute for a proper treatment of such subjects but I feel that this book is not the place for yet another repetition of well-worn material which is readily available in existing texts.

There now follows a brief guide to the contents. When the Laplace transform is taken of a system of ordinary linear differential equations with constant coefficients and the resulting equations expressed in matrix-vector form, the matrices involved have elements which are polynomials. Chapter 1 begins with a short summary of the basic properties of polynomial matrices, and then presents some recent results by the author. In Section 2 some necessary and sufficient conditions are derived for two square polynomial matrices to have relatively prime determinants, and these are followed by theorems concerning the invariant factors of the two matrices. The chapter continues with an introduction to Rosenbrock's theory of linear multivariable systems (from which the preceding results arose), and lays particular emphasis on the concept of primeness of two polynomial matrices. An application of polynomial matrices to linear programming is also given.

During the last century a number of useful results on the qualitative analysis of polynomials were discovered, largely in terms of determinants. Topics of interest include the decision whether two polynomials have a greatest common divisor and if so what this divisor is, and examination of a single polynomial for repeated zeros. Although still of considerable interest these older results involving resultants, discriminants and bigradients are not found in many present-day textbooks, and statements of the main theorems are therefore given in the first part of Chapter 2. In the second section some recent work is presented whereby these classical results can be expressed rather more simply in terms of matrices, and this leads to some valuable extensions. The problems of determining how many zeros of a given complex polynomial lie in a given half-plane or inside the unit circle are of long-standing interest in control theory and are closely related to those in the preceding sections. Here, too, it is possible to formulate the well-known results of Hermite, Cohn, Liénard and Chipart in matrix terms.

Matrices whose elements are rational functions play a fundamental role in linear multivariable control theory, and these are dealt with in Chapter 3. After a brief introduction the realization problem of finding linear systems which correspond to a given transfer function matrix is considered, and this area is also explored in the context of Rosenbrock's theory introduced in Chapter 1. Positive-real matrices and the related spectral factorization problem are defined and some basic theorems are given.

The stability of linear constant-coefficient differential equations is a well-worn topic but when viewed in matrix terms the application of Liapunov theory produces some new and interesting results which are set out in the first part of Chapter 4. This leads to consideration of the location of characteristic roots of complex matrices and some extensions of the preceding results are derived. Related situations, arising from linear economic theory in which only the signs of the matrix elements are known, are also discussed.

In linear control theory where the object is to choose the control so as to minimize a quadratic performance index, the solution depends upon a matrix differential equation of Riccati type. In Chapter 5, before looking at this particular case, some theorems are given on general matrix Riccati equations and it is then shown how these specialize for the equation which arises in control. If the optimization is carried out over an infinite time period a purely algebraic equation results, and some interesting material on this and more general quadratic matrix equations closes the chapter.

Chapter 6 is concerned with generalizations of the idea of inverse to cover singular or rectangular matrices. This has found application in many areas of linear mathematics and is especially related to solution of systems of linear algebraic equations. The best known generalized inverse is that associated with Moore and Penrose, and short accounts of basic properties of this have appeared in a few recent textbooks, but the discussion which begins the chapter is a rather more extended one and is followed by a section in which other types of generalized inverse are defined and investigated. Again, it is possible to show an application to linear programming.

The theme of linear equations is continued in Chapter 7, but from the viewpoint of seeking solutions in integers, as the practical application of this is to integer programming. In matrix terms this involves concepts of unimodularity, but much of the material which has appeared in the literature is not too easy to follow (for example, often being expressed in graph-theoretic terms). In order to overcome this difficulty the account in Section 1 is deliberately expressed in a very simple form. The relationship in linear programming to transportation problems, whose solutions always come out in integers, is explained and a special type of equivalence relation is defined. It is interesting that the matrices having integer elements which appear in this final chapter share a fundamental property with the polynomial matrices of Chapter 1, the connecting link being the theorem produced by H. J. S. Smith in 1861.

Reading through the preceding summary may tempt the reader to protest that he is to be confronted with a hotch-potch of miscellaneous results! I hope that this proves to be a harsh judgment and that the variety and scope of material to be found in the following pages will in fact be one of its main attractions. The selection of topics has, of course, been influenced by my own personal interests, and I can only apologize if the reader feels that some areas have been treated too fully and others neglected. I must also beg his indulgence if he feels that too much of my own work in matrix theory has been included. Nevertheless, I have done my best to include results which seem likely to be of lasting value and interest, and any blame for errors or omissions is entirely my own.

Finally, I have made only brief mention of numerical aspects as these lie outside my field of competence. However, on many occasions whilst preparing material for this book I have felt that there is a need for a text devoted to description and critical analysis of numerical algorithms in control theory, and it is to be hoped that this gap in the literature will soon be filled.

It is a pleasure to thank Professor H. H. Rosenbrock for several valuable comments and suggestions. I am also grateful to Professor J. B. Helliwell for making available resources of the University of Bradford, and to Mrs Margaret Balmforth and Mrs June Russell who between them made an excellent job of typing the manuscript. Finally, I should like to thank my publishers, Van Nostrand Reinhold Company Ltd, for their helpful assistance throughout the project.

STEPHEN BARNETT
Bradford, Yorkshire
December 1970

List of Symbols

$\Rightarrow$	implies
$\Leftrightarrow$	if and only if
$x \in S$	x is a member of the set S
$\bar{z}$	conjugate of the complex number z
$\lvert z \rvert = (\bar{z}z)^{\frac{1}{2}}$	modulus of z
Re (z)	real part of z
$A = [a_{ij}]$	matrix having a_{ij} as element in row i, column j
$b = [b_i]$	vector having b_i as ith component
$A^T = [a_{ji}]$	transpose of A
A^{-1}	inverse of a nonsingular square matrix A
$\bar{A} = [\bar{a}_{ij}]$	complex conjugate of A
$A^* = [\bar{a}_{ji}]$	conjugate transpose of A
det A or $\lvert A \rvert$	determinant of A
Adj A	adjoint of A
tr $A = \Sigma a_{ii}$	trace of A

$\lvert\lvert A \rvert\rvert = (\sum_i \sum_j \lvert a_{ij} \rvert^2)^{1/2} = \text{tr } A^*A$ Euclidean norm of A

$\lvert\lvert b \rvert\rvert = (\Sigma \lvert b_i \rvert^2)^{1/2}$ Euclidean norm of b

$A = \text{diag } [k_1, k_2, \ldots, k_n]$ $n \times n$ matrix A having $a_{ii} = k_i$ and $a_{ij} = 0$, $i \neq j$

$I_n = \text{diag } [1, 1, \ldots, 1]$ unit matrix of order n

A^+	Moore–Penrose generalized inverse of A
$\delta\{A(\lambda)\}$	degree of the polynomial matrix $A(\lambda)$
$\delta[A(\lambda)]$	degree of the rational matrix $A(\lambda)$
$A \geqslant B$	$a_{ij} \geqslant b_{ij}$, all possible i and j
$A \otimes B$	Kronecker product of A and B
$\dot{x}(t)$	derivative of $x(t)$ with respect to t
$a'(\lambda)$	derivative of $a(\lambda)$ with respect to λ

$$\text{sgn } k = \begin{cases} +1, & k > 0 \\ -1, & k < 0 \\ 0, & k = 0 \end{cases} \quad \text{signum function}$$

Contents

1

Polynomial Matrices

1.1 Definitions and basic properties

We begin with a brief summary of some properties of polynomial matrices which will be needed in the remainder of the chapter. No attempt has been made to give a complete account since most of the results quoted are well known and may be found in the majority of the textbooks referred to in Appendix 1. These should be consulted for proofs and further details and developments.

A rectangular polynomial matrix $A(\lambda)$ is an $m \times n$ matrix whose elements are polynomials in λ, the coefficients of the polynomials belonging to a field of numbers which in this chapter will be taken to be the field of complex numbers. If the highest degree of the elements of $A(\lambda)$ is N then we can write

$$A(\lambda) = A_0\lambda^N + A_1\lambda^{N-1} + A_2\lambda^{N-2} + \cdots + A_N, \qquad (1.1.1)$$

where the A_i are $m \times n$ matrices with complex elements, and N is said to be the *degree* of A, written $\delta\{A(\lambda)\}$. Addition, subtraction and multiplication of polynomial matrices (if possible) are defined in an obvious and straightforward fashion. For example, if $B(\lambda)$ is a second matrix having dimensions $n \times p$ and degree M then

$$A(\lambda)B(\lambda) = \sum_{i=0}^{M+N} C_i\lambda^{M+N-i}, \qquad \text{where } C_i = \sum_{j+k=i} A_jB_k,$$

but notice that $\delta(AB) \leqslant \delta(A) + \delta(B)$, for $C_0 = A_0B_0$ may be the $m \times p$ zero matrix.

When $A(\lambda)$ is $n \times n$ then its determinant is also a polynomial and clearly has degree less than or equal to nN. However Belevitch [6] has shown that

$$\delta\{\det A(\lambda)\} \leqslant \sum_{i=0}^{N-1} r_i,$$

where $r_i = \text{rank}\,[A_0, A_1, \ldots, A_i]$, with equality holding if all A_i are real, symmetric and positive semidefinite. Since the coefficient of λ^{nN} in $\det A(\lambda)$ is $\det A_0$, $\delta\{\det A(\lambda)\} = nN$ only if A_0 is nonsingular, in which case $A(\lambda)$ is termed *regular*. In particular, if either of $A(\lambda)$ or $B(\lambda)$ is regular then, provided $A(\lambda)B(\lambda)$ exists, $\delta(AB) = \delta(A) + \delta(B)$.

The inverse of a square matrix $A(\lambda)$ will not in general be a polynomial matrix, since $A^{-1}(\lambda) = \text{Adj } A(\lambda)/\det A(\lambda)$; only if $\det A(\lambda)$ is a nonzero scalar will $A^{-1}(\lambda)$ be a polynomial matrix (this follows from $\det A \det A^{-1} = 1$), and $A(\lambda)$ is then called *invertible*. Unfortunately some authors (for example [16]) apply the term 'regular' in these circumstances (thus giving it quite a different meaning to that of the previous paragraph) but our usage is a more natural one being in line with terminology for matrices with elements belonging to a number field.† When $A(\lambda)$ is invertible clearly $\det A(\lambda) = \det A_N$, $\det A_0 = 0$ and $\delta\{A^{-1}(\lambda)\} = \delta\{\text{Adj } A(\lambda)\} \leqslant (n-1)N$.

Division involving two square matrices of the same order is dealt with by the following theorems.

THEOREM 1.1 If $A(\lambda)$ and $B(\lambda)$ are of order n and $A(\lambda)$ is regular then there exists exactly one pair of matrices $Q_1(\lambda)$, $R_1(\lambda)$ such that

$$B(\lambda) = Q_1(\lambda)A(\lambda) + R_1(\lambda)$$

and $\delta(R_1) < \delta(A)$; and exactly one pair $Q_2(\lambda)$, $R_2(\lambda)$ such that

$$B(\lambda) = A(\lambda)Q_2(\lambda) + R_2(\lambda)$$

and

$$\delta(R_2) < \delta(A).$$

For a proof, see, for example, [10]. The matrices $Q_1(\lambda)$and $R_1(\lambda)$ are called the *right quotient* and *right remainder* of $B(\lambda)$ on division by $A(\lambda)$; similarly, $Q_2(\lambda)$ and $R_2(\lambda)$ are left quotient and remainder respectively. Notice that each quotient is either identically zero or has degree $\delta\{B(\lambda)\} - \delta\{A(\lambda)\}$.

When the divisor $A(\lambda)$ is not regular there is no longer a unique quotient and remainder (with degree less than that of $A(\lambda)$). It is easy to show, however, that division is still possible:

THEOREM 1.2 If $A(\lambda)$ is nonsingular (i.e. $\det A(\lambda) \not\equiv 0$) then there exists exactly one pair of matrices $Q_3(\lambda)$, $R_3(\lambda)$ such that

$$B(\lambda) = Q_3(\lambda)A(\lambda) + R_3(\lambda)$$

and

$$\delta(R_3 \text{ Adj } A) < \delta(\det A), \qquad \delta(R_3) < \delta(A).$$

Proof. Let $a(\lambda) = \det A(\lambda)$. By Theorem 1.1, there exists a unique pair of matrices $Q^1(\lambda)$, $R^1(\lambda)$ such that

$$B(\lambda)\{\text{Adj } A(\lambda)\} = Q^1(\lambda)a(\lambda) + R^1(\lambda),$$

† *The term 'proper' is sometimes used instead of 'regular' when* $\det A_0 \neq 0$ [10; 22, *p.* 45].

with

$$\delta\{R^1(\lambda)\} < \delta\{a(\lambda)\}.$$

Therefore

$$B(\text{Adj } A)A = Q^1Aa + R^1A,$$

so

$$B = Q^1A + R^1A/a.$$

Thus

$$Q_3(\lambda) = Q^1(\lambda),\ R_3(\lambda) = R^1(\lambda)A(\lambda)/a(\lambda)$$

and

$$\delta\{R_3(\lambda)\} \leqslant \delta\{R^1(\lambda)\} + \delta\{A(\lambda)\} - \delta\{a(\lambda)\} < \delta\{A(\lambda)\}.$$

Also, $R_3(\lambda)$ Adj $A(\lambda) = R^1(\lambda)$, so $\delta\{R_3(\lambda) \text{ Adj } A(\lambda)\} < \delta\{a(\lambda)\}$. Notice that $Q_3 = 0$ if and only if $\delta(B \text{ Adj } A) < \delta(a)$; otherwise

$$\delta(Q_3) = \delta(B \text{ Adj } A) - \delta(a).$$

That division is not unique is easily seen, since

$$B = (Q_3 + K)A + (R_3 - KA)$$

where $K(\lambda)$ is any matrix such that $\delta\{K(\lambda)A(\lambda)\} < \delta\{A(\lambda)\}$.

Flood [13] gives further discussion of the case when the divisor $A(\lambda)$ is not regular but is nonsingular and shows how all divisions of $B(\lambda)$ by $A(\lambda)$ may be found, together with various other related results. Unfortunately his notation is rather cumbersome and awkward to follow.

The *rank* of a polynomial matrix is the order of the largest minor which is not identically zero. Thus, for example, the rank of a regular matrix is equal to its order. We define the following *elementary operations* on a rectangular polynomial matrix which leave its rank unaltered:

(i) multiply any line by a nonzero scalar
(ii) add to any line any other line multiplied by an arbitrary polynomial
(iii) interchange any two lines
('line' stands for either row or column of the matrix).

When any one of these operations is performed on a unit matrix I_n, the resulting matrix (which has a determinant independent of λ) is called an *elementary matrix*.

For example when $n = 3$, typical elementary matrices formed by operating on the columns of I_3 are

$$\begin{bmatrix} 1 & 0 & 0 \\ 0 & a & 0 \\ 0 & 0 & 1 \end{bmatrix}, \quad \begin{bmatrix} 1 & 0 & 0 \\ 0 & 1 & b(\lambda) \\ 0 & 0 & 1 \end{bmatrix}, \quad \begin{bmatrix} 1 & 0 & 0 \\ 0 & 0 & 1 \\ 0 & 1 & 0 \end{bmatrix}.$$

It is easy to verify that when any $m \times 3$ matrix is multiplied on the right by one of these elementary matrices, this results in the corresponding elementary operation being applied to its columns. Similarly in general: to bring about an elementary transformation on any given matrix $A(\lambda)$, the same transformation is first applied to the unit matrix of appropriate order and this elementary matrix then postmultiplies $A(\lambda)$ for column operations and premultiplies for row operations.

Two polynomial matrices are said to be *equivalent* or to be related by an *equivalence transformation* if it is possible to pass from one to the other by a sequence of elementary operations. It can be shown that any invertible polynomial matrix can be expressed as a product of elementary matrices, and it therefore follows that:

THEOREM 1.3 $A(\lambda)$ and $B(\lambda)$ are equivalent $\{A(\lambda) \sim B(\lambda)\}$ if and only if there exist invertible matrices $P(\lambda)$, $Q(\lambda)$ such that

$$A(\lambda) = P(\lambda)B(\lambda)Q(\lambda).$$

We can now go on to another important aspect of equivalence. Let $A(\lambda)$ be an $m \times n$ matrix of rank r and let $d_j(\lambda)$ stand for the greatest common divisor (g.c.d.) of all jth order minors of $A(\lambda)$ ($j = 1, 2, \ldots, r$). The polynomials $d_j(\lambda)$ are called the *determinantal divisors* and we can take each to be *monic* (i.e. with the highest degree term having coefficient unity) without loss of generality. It is easily seen that d_j is divisible by d_{j-1} (taking $d_0 = 1$), so that the quotients $i_j(\lambda) = d_j/d_{j-1}$ are also monic polynomials. These are called the *invariant polynomials* or *invariant factors* of $A(\lambda)$ because they are unaltered by equivalence transformations on $A(\lambda)$, as expressed in the following theorems.

THEOREM 1.4 $A(\lambda)$ and $B(\lambda)$ are equivalent if and only if they have the same invariant factors.

THEOREM 1.5 A matrix $A(\lambda)$ of rank r is equivalent to

$$\text{diag}\,[i_1(\lambda), i_2(\lambda), \ldots, i_r(\lambda), 0, \ldots, 0],$$

the so-called *Smith canonical* or *normal* form of a polynomial matrix. Notice that $i_j(\lambda)$ is a factor of $i_{j+1}(\lambda)$ for all j.

In fact Smith [41] developed the canonical form for matrices with integer elements, but since polynomials over a field and the set of integers are both examples of a principal ideal domain (see Appendix 1) the method of proof is easily modified. Although the proof is a constructive one (see for example [14, p. 137]), computation of the Smith form is in general a lengthy process, but Dewey [12] has outlined a computer programme which is reported to have worked successfully for fourth order matrices of high degree. At present, however, nothing is known about the degree of the matrices $P(\lambda)$ and $Q(\lambda)$ in Theorem 1.3 when $B(\lambda)$ is the Smith form of $A(\lambda)$, except that in the special case when

$A(\lambda)$ and $B(\lambda)$ are both regular and have degree one then $P(\lambda)$ and $Q(\lambda)$ can be taken independent of λ.

Since all the polynomials occurring here have complex coefficients it follows that each $i_j(\lambda)$ can be written as a product of linear factors

$$i_j(\lambda) = (\lambda - \alpha_1)^{k_{j1}}(\lambda - \alpha_2)^{k_{j2}} \cdots (\lambda - \alpha_s)^{k_{js}} \tag{1.1.2}$$

where each k_{js} is a nonnegative integer and $(\lambda - \alpha_1), \ldots, (\lambda - \alpha_s)$ are the distinct linear factors of $i_r(\lambda)$; notice that $k_{rl} > 0, l = 1, 2, \ldots, s$. All those factors $(\lambda - \alpha_l)^{k_{jl}}$ which are not unity are called the *elementary divisors* of $A(\lambda)$. Since i_j divides i_{j+1}, it follows that

$$\cdots \leqslant k_{j\theta} \leqslant k_{j+1,\theta} \leqslant k_{j+2,\theta} \leqslant \cdots \text{etc.}$$

In the case when $A(\lambda)$ is square of order n and degree N and is regular, clearly

$$\sum_j \sum_l k_{jl} = nN,$$

since $\det A(\lambda) = i_1(\lambda)i_2(\lambda) \cdots i_n(\lambda)$ and $\delta(\det A) = nN$. The numbers $\lambda_1, \lambda_2, \ldots, \lambda_s$ are the roots of $\det A(\lambda) = 0$, and have been termed the *latent roots* of $A(\lambda)$ by Lancaster [21]. His book is concerned with problems of vibrating systems and contains a chapter on numerical methods for computing latent roots and associated latent vectors. We refer the reader to this text [21] for further details and a number of other interesting results on polynomial matrices.

A polynomial matrix of special importance is the characteristic matrix $\lambda I_n - K$ of an $n \times n$ matrix K with complex elements. If

$$\det(\lambda I_n - K) = (\lambda - \lambda_1)^{k_1}(\lambda - \lambda_2)^{k_2} \cdots (\lambda - \lambda_r)^{k_r},$$

where the λ_i are all distinct, then $k_1 + k_2 + \cdots + k_r = n$, since $\lambda I_n - K$ is regular. These latent roots are called the *characteristic roots* (or commonly, the *eigenvalues*) of K and any nonzero vector satisfying

$$(\lambda_i I_n - K)x_i = 0$$

is a *characteristic vector* associated with λ_i. If K has fewer than n linearly independent characteristic vectors it is said to be *defective*. If K is not defective (for example, if all its characteristic roots $\lambda_1, \ldots, \lambda_n$ are distinct) then it is similar to a diagonal matrix. More generally:

THEOREM 1.6 K is similar to the *Jordan form*

$$\operatorname{diag}[\mathcal{J}_{m_1}(\lambda_1), \mathcal{J}_{m_2}(\lambda_1), \ldots, \mathcal{J}_{m_s}(\lambda_1), \mathcal{J}_{n_1}(\lambda_2), \ldots, \mathcal{J}_{q_u}(\lambda_r)]$$

where

$$\mathcal{J}_k(\lambda) = \begin{bmatrix} \lambda & 1 & 0 & 0 & \cdots & \cdot & \cdot \\ 0 & \lambda & 1 & 0 & \cdots & \cdot & \cdot \\ 0 & 0 & \lambda & 1 & \cdots & \cdot & \cdot \\ \cdot & \cdot & \cdot & \cdot & \cdots & \cdot & \cdot \\ \cdot & \cdot & \cdot & \cdot & \cdots & \lambda & 1 \\ \cdot & \cdot & \cdot & \cdot & \cdots & 0 & \lambda \end{bmatrix}$$

is a *Jordan block* of order k, and the elementary divisors of $\lambda I_n - K$ are

$$(\lambda - \lambda_1)^{m_1}, (\lambda - \lambda_1)^{m_2}, \ldots, (\lambda - \lambda_1)^{m_s}, (\lambda - \lambda_2)^{n_1}, \ldots, (\lambda - \lambda_r)^{q_u},$$

so that $m_1 + m_2 + \cdots + m_s = k_1$, etc.

For a proof, see [14] or [10].

If there is more than one Jordan block associated with λ_i for some i, K is termed *derogatory*. Thus a *nonderogatory* matrix has only one Jordan block associated with each distinct λ_i (see Appendix 1 for an alternative definition).

The only other theorem we shall need for later development closes this section.

THEOREM 1.7 If $A(\lambda) = I_n\lambda^N + A_1\lambda^{N-1} + \cdots + A_N$, then $\det(\lambda I_{nN} - C) \equiv \det A(\lambda)$, where C is the $(nN) \times (nN)$ matrix

$$\begin{bmatrix} 0 & 0 & \ldots & . & -A_N \\ I_n & 0 & \ldots & . & -A_{N-1} \\ 0 & I_n & \ldots & . & -A_{N-2} \\ . & . & \ldots & . & . \\ 0 & 0 & \ldots & I_n & -A_1 \end{bmatrix}. \tag{1.1.3}$$

Proof. This readily follows by analogy with the case when $n = 1$ (when C is a companion matrix of the polynomial $A(\lambda)$, see Appendix 1), for example by induction on N, and is left as an exercise for the reader.

Incidentally, there is no loss of generality caused by taking A_0 to be the unit matrix in Theorem 1.7 provided $A(\lambda)$ is regular.

EXERCISES

1.1.1 Show that if $B(\lambda)$ is a matrix of order n and degree k, then the right and left remainders on division by $\lambda I_n - A$ (where A is a constant matrix) are $B_0A^k + B_1A^{k-1} + \cdots + B_k$ and $A^kB_0 + A^{k-1}B_1 + \cdots + B_k$ respectively.

1.1.2 Let $R(\lambda) = I_n + R_1\lambda$ with $R_1^n = 0$. By considering the product $R(\lambda)T(\lambda)$, where $T(\lambda) = T_0\lambda^{n-1} + \cdots + T_{n-2}\lambda + I_n$ show that $R(\lambda)$ is invertible.

1.1.3 Show that the product of two invertible polynomial matrices is also an invertible matrix. Hence show, using the previous exercise, that $R_1^2\lambda^2 + 2R_1\lambda + I_n$ and $R_1S_1\lambda^2 + (R_1 + S_1)\lambda + I_n$ are invertible matrices, where R_1 and S_1 are constant matrices such that $R_1^n = S_1^n = 0$.

1.1.4 If $R(\lambda)$ is an invertible matrix show that $dR/d\lambda$ and $dR^{-1}/d\lambda$ are equivalent.

1.1.5 [2] Show that if $N(\lambda) = N_0\lambda^p + N_1\lambda^{p-1} + \cdots + N_p$, $(p \geqslant 2)$ is a matrix with inverse $\nu(\lambda) = \nu_0\lambda^q + \cdots + \nu_q$, $(q \geqslant 2)$ then both $N_0\lambda + N_1$ and $\nu_0\lambda + \nu_1$ are invertible provided N_1 and ν_1 are nonsingular.

1.1.6 Verify that

$$\det\left(\begin{bmatrix} 0 & 0 & I_n \\ 0 & I_n & A_1 \\ I_n & A_1 & A_2 \end{bmatrix}\lambda + \begin{bmatrix} 0 & -I_n & 0 \\ -I_n & -A_1 & 0 \\ 0 & 0 & A_3 \end{bmatrix}\right)$$
$$= (-1)^n \det(I_n\lambda^3 + A_1\lambda^2 + A_2\lambda + A_3).$$

This is easily generalized for matrices of higher degree [21, p. 58].

1.1.7 If A_0 in (1.1.1) is singular but A_N is nonsingular, obtain a result corresponding to Theorem 1.7.

1.1.8 If $n = 1$ in Theorem 1.7, C is called a *companion matrix* of the (scalar) polynomial $a(\lambda)$ (see Appendix 1). Show that $\lambda I_N - C$ is nonderogatory by reducing it to diag $[1, 1, \ldots, 1, a(\lambda)]$ by means of the following equivalence transformations: add λ times the second row onto the first, then interchange the first and second rows. Add suitable multiples of the first column onto the second and last columns to reduce the rest of the first row to zero. Repeat with the second and third rows, etc.

1.2 Matrices having relatively prime determinants

A fundamental theorem concerning polynomials runs as follows: two polynomials $a(\lambda)$ and $b(\lambda)$ are relatively prime (i.e. their greatest common divisor is independent of λ) if and only if there exist a unique pair of polynomials $x(\lambda)$, $y(\lambda)$ with $\delta x < \delta b$ and $\delta y < \delta a$ such that $ax + by = 1$ (see Appendix 1).

Our aim in this section is to show how this result can be generalized to two regular polynomial matrices

$$A(\lambda) = I_n\lambda^N + A_1\lambda^{N-1} + \cdots + A_N \tag{1.2.1}$$
$$B(\lambda) = I_m\lambda^M + B_1\lambda^{M-1} + \cdots + B_M. \tag{1.2.2}$$

The theorem we shall prove in fact turns out to be a condition for relative primeness of the determinants of $A(\lambda)$ and $B(\lambda)$, and may be stated as follows:

THEOREM 1.8 [4] The determinants of $A(\lambda)$ and $B(\lambda)$ in (1.2.1) and (1.2.2) are relatively prime if and only if the equation

$$A(\lambda)X(\lambda) + Y(\lambda)B(\lambda) = E \tag{1.2.3}$$

for a given $n \times m$ matrix E with constant elements, has a unique solution $X(\lambda)$, $Y(\lambda)$ with $\delta\{X(\lambda)\} < \delta\{B(\lambda)\}$ and $\delta\{Y(\lambda)\} < \delta\{A(\lambda)\}$.

The result for polynomials is therefore just a special case of Theorem 1.8 when $n = m = 1$ so that the proof given below incidentally provides a new proof for this simple case. Notice that when we say E has constant elements, we mean elements belonging to the same field of numbers as the coefficients of the polynomials, which we have taken as the field of complex numbers throughout this chapter.

Before proving the theorem we shall need a preliminary lemma, which is of considerable importance in its own right and will be required again in later chapters.

THEOREM 1.9 The equation

$$\alpha\gamma - \gamma\beta = \delta, \tag{1.2.4}$$

where α, β, δ are matrices with constant elements and dimensions $s \times s$, $t \times t$ and $s \times t$ respectively, has a unique solution γ if and only if α and β have no common characteristic roots.

Proof. The simplest approach, although not that followed by Gantmacher [14], for example, is to write (1.2.4) in the form

$$(\alpha \otimes I_t - I_s \otimes \beta^T)\gamma' = \delta', \tag{1.2.5}$$

where $\otimes$ denotes Kronecker product and γ' and δ' are column vectors of st elements formed from the rows of γ, δ respectively taken in order (see Appendix 1). Equation (1.2.4) has a unique solution if and only if the matrix on the left in (1.2.5) is nonsingular, and this matrix has characteristic roots $\lambda_i - \mu_j$, λ_i and μ_j being the roots of α and β respectively.

We can now give the

Proof of Theorem 1.8. Let the remaining matrices in (1.2.3) be written

$$X(\lambda) = X_0\lambda^{M-1} + X_1\lambda^{M-2} + \cdots + X_{M-1} \tag{1.2.6}$$

and

$$Y(\lambda) = Y_0\lambda^{N-1} + Y_1\lambda^{N-2} + \cdots + Y_{N-1}, \tag{1.2.7}$$

where X_i and Y_i are $n \times m$ matrices with complex elements.

Further, let

$$D = \begin{bmatrix} 0 & I_m & 0 & \cdots & \cdot \\ 0 & 0 & I_m & \cdots & \cdot \\ \cdot & \cdot & \cdot & \cdots & \cdot \\ \cdot & \cdot & \cdot & \cdots & I_m \\ -B_M & -B_{M-1} & -B_{M-2} & \cdots & -B_1 \end{bmatrix} \tag{1.2.8}$$

be the $(mM) \times (mM)$ matrix obtained from $B(\lambda)$ in a similar way to that in which C is defined in (1.1.3), so that as in Theorem 1.7,

$\det(\lambda I_{mM} - D) \equiv \det B(\lambda)$. The basis of the proof is to establish a relationship between the solution (1.2.6), (1.2.7) of equation (1.2.3) and the solution for the matrix F of the equation

$$CF - FD = G. \tag{1.2.9}$$

In (1.2.9) both F and G are $(nN) \times (mM)$ constant matrices partitioned into $n \times m$ submatrices F_{ij}, G_{ij} respectively ($i = 1, 2, \ldots, N$; $j = 1, 2, \ldots, M$).

First, by virtue of (1.2.1), (1.2.2), (1.2.6), and (1.2.7), the coefficient of λ^k on the left side of (1.2.3) is given by

$$\left.\begin{aligned} T_k &= \sum_{i+j=l} (A_i X_j + Y_i B_j) \qquad (k = 0, 1, 2, \ldots, M + N - 2; \\ &\qquad\qquad\qquad\qquad\qquad l = M + N - 1 - k) \\ &\text{and} \\ T_{M+N-1} &= X_0 + Y_0. \end{aligned}\right\} \tag{1.2.10}$$

Next, using (1.1.3) and (1.2.8) it follows from equation (1.2.9) that

$$G_{ij} = F_{i-1,j} - A_{N-i+1}F_{Nj} - F_{i,j-1} + F_{iM}B_{M-j+1} \tag{1.2.11}$$
$$(i = 1, 2, \ldots, N; j = 1, 2, \ldots, M)$$

with $F_{0j} = 0 = F_{i0}$. Setting

$$X_{M-j} = -F_{Nj}, \; Y_{N-i} = F_{iM} \tag{1.2.12}$$
$$(i = 1, 2, \ldots, N; j = 1, 2, \ldots, M)$$

it then follows from (1.2.10), (1.2.11) and (1.2.12), after a little manipulation, that

$$T_k = \sum_{i+j=k+2} G_{ij} \qquad (k = 0, 1, 2, \ldots, M + N - 2)$$

and

$$T_{M+N-1} = 0.$$

Thus if F is any solution of equation (1.2.9) with $G_{11} \equiv E$, all other $G_{ij} \equiv 0$, then $X(\lambda)$, $Y(\lambda)$ given by (1.2.6), (1.2.7) and (1.2.12) is a solution of (1.2.3), for then $T_0 = E$ and all other $T_k \equiv 0$; in other words the left side of (1.2.3) reduces to E as required. Conversely, if $X(\lambda)$, $Y(\lambda)$ is any solution of (1.2.3) then F_{ij} given by (1.2.12) and (1.2.11) is a solution of (1.2.9) with $G_{11} = E$, all other $G_{ij} \equiv 0$. Hence the solution of (1.2.3) is unique if and only if the solution of (1.2.9) (for the stated choice of G) is unique. But by Theorem 1.9, this latter holds if and only if C and D have no common characteristic roots; that is, by Theorem 1.7, if and only if $\det A(\lambda)$ and $\det B(\lambda)$ are relatively prime.

Theorem 1.8 still applies if E is replaced by a polynomial matrix of degree less than or equal to $\delta(A) + \delta(B) - 2$, and thus provides a sufficient condition for a solution of (1.2.3) to exist in such circumstances. Specifically, if $E(\lambda) = E_0\lambda^{M+N-2} + \cdots + E_{M+N-2}$, the only alteration necessary in the proof is that the G_{ij} and E_j must satisfy

$$\sum_{i+j=k+2} G_{ij} = E_{M+N-k-2} \qquad (k = 0, 1, 2, \ldots, M + N - 2)$$

and this leads to no inconsistencies.

A theorem on the existence of a solution to a more general form of equation (1.2.3) has been given by Roth [38] and may be stated here without proof:

THEOREM 1.10 A necessary and sufficient condition that there exists a solution $X(\lambda)$, $Y(\lambda)$ to the equation

$$A(\lambda)X(\lambda) + Y(\lambda)B(\lambda) = E(\lambda),$$

where $A(\lambda)$, $B(\lambda)$ and $E(\lambda)$ are each of order n, is that the matrices

$$\begin{bmatrix} A(\lambda) & E(\lambda) \\ 0 & B(\lambda) \end{bmatrix} \quad \text{and} \quad \begin{bmatrix} A(\lambda) & 0 \\ 0 & B(\lambda) \end{bmatrix}$$

be equivalent.

Notice that there is no restriction on the degree of $E(\lambda)$ and that the theorem also holds if the matrices are rectangular. Jones [17] has also derived some results for a slightly more general equation.

An interesting alternative way of expressing the condition of Theorem 1.8 is obtained by writing the equations

$$\left.\begin{array}{ll} T_k = 0 & (k = 1, 2, \ldots, M + N - 1) \\ T_0 = E & \end{array}\right\} \qquad (1.2.13)$$

obtained from (1.2.10), in terms of column vectors x_i, y_j of the rows of X_i, Y_j respectively taken in order (as in the proof of Theorem 1.9). We obtain

$$\begin{aligned} \sum_{i+j=l} [(A_i \otimes I_m)x_j + (I_n \otimes B_j^T)y_i] &= 0 \qquad (k = 1, 2, \ldots, M + N - 2) \\ &= e \qquad (k = 0) \qquad (1.2.14) \\ (I_n \otimes I_m)x_0 + (I_n \otimes I_m)y_0 &= 0, \end{aligned}$$

e being the column vector obtained from the matrix E. The condition for (1.2.3) to have a unique solution is the same as the condition that the equations (1.2.14) have a unique solution, namely that the matrix of coefficients on the left of (1.2.14) be nonsingular. After rearrangement and interchange of block rows and columns (as in the scalar case [7, p. 195]; see also Chapter 2, Section 1) we obtain

THEOREM 1.11 The determinant

$$\begin{vmatrix} \tau_0 & \tau_1 & \cdots & \cdot & \cdot & \tau_N & 0 & 0 & \cdot & 0 \\ 0 & \tau_0 & \cdots & \cdot & \cdot & \tau_{N-1} & \tau_N & 0 & \cdot & 0 \\ \cdot & \cdot & \cdots & \cdot & \cdot & \cdot & \cdot & \cdot & \cdot & \cdot \\ \cdot & \cdot & \cdots & \cdot & \cdot & \cdot & \cdot & \cdot & \tau_{N-1} & \tau_N \\ \cdot & \cdot & \cdots & \cdot & \mu_0 & \mu_1 & \cdot & \cdot & \mu_{M-1} & \mu_M \\ \cdot & \cdot & \cdots & \mu_0 & \mu_1 & \mu_2 & \cdot & \cdot & \mu_M & 0 \\ \cdot & \cdot & \cdots & \cdot & \cdot & \cdot & \cdot & \cdot & \cdot & \cdot \\ \mu_0 & \mu_1 & \cdots & \cdot & \cdot & \cdot & \cdot & \cdot & \cdot & \cdot \end{vmatrix}, \qquad (1.2.15)$$

where

$$\tau_i = A_i \otimes I_m \qquad (i = 0, 1, \ldots, N;\ A_0 \equiv I_n)$$
$$\mu_j = I_n \otimes B_j \qquad (j = 0, 1, \ldots, M;\ B_0 \equiv I_m),$$

does not vanish if and only if det $A(\lambda)$ and det $B(\lambda)$ are relatively prime.

Notice that since det $B^T(\lambda) \equiv$ det $B(\lambda)$, B_j^T in (1.2.14) has been replaced by B_j in (1.2.15).

The determinant (1.2.15) can in fact be reduced to one of much smaller order by noting that det $A(\lambda)$ and det $B(\lambda)$ are relatively prime if and only if

$$\det (C \otimes I_{mM} - I_{nN} \otimes D) \neq 0 \qquad (1.2.16)$$

(this follows easily from Theorems 1.7 and 1.9). The determinant in (1.2.16) is thus equivalent to that in (1.2.15) but because of its form:

$$\begin{vmatrix} -I_n \otimes D & 0 & \cdots & \cdot & -A_N \otimes I_{mM} \\ I_n \otimes I_{mM} & -I_n \otimes D & \cdots & \cdot & -A_{N-1} \otimes I_{mM} \\ 0 & I_n \otimes I_{mM} & \cdots & \cdot & -A_{N-2} \otimes I_{mM} \\ \cdot & \cdot & \cdots & \cdot & \cdot \\ \cdot & \cdot & \cdots & I_n \otimes I_{mM} & \begin{pmatrix} -I_n \otimes D \\ -A_1 \otimes I_{mM} \end{pmatrix} \end{vmatrix}$$

it can be reduced by adding $(I_n \otimes D)$ times the second block row onto the first block row, and then expanding by the first block column. In the resulting determinant, add $(I_n \otimes D^2)$ times the second block row to the first and again expand by the first column. The process is repeated to end up with

$$(-1)^{N-2} \begin{vmatrix} (-I_n \otimes D^{N-1}) & (-A_N \otimes I_{mN} - A_{N-1} \otimes D - \cdots - A_2 \otimes D^{N-2}) \\ (I_n \otimes I_{mN}) & (-A_1 \otimes I_{mM} - I_n \otimes D) \end{vmatrix}$$
$$= (-1)^{N-2} \det (I_n \otimes D^N + A_1 \otimes D^{N-1} + \cdots + A_N \otimes I_{mM}),$$

a determinant of order mnM. We have thus proved

THEOREM 1.12 A necessary and sufficient condition for det $A(\lambda)$

and det $B(\lambda)$ to be relatively prime is that

$$\det(I_n \otimes D^N + A_1 \otimes D^{N-1} + A_2 \otimes D^{N-2} + \cdots + A_N \otimes I_{mM}) \neq 0. \quad (1.2.17)$$

Clearly, by interchanging C and D, (1.2.17) is equivalent to

$$\det(I_m \otimes C^M + B_1 \otimes C^{M-1} + B_2 \otimes C^{M-2} + \cdots + B_M \otimes I_{nN}) \neq 0, \quad (1.2.18)$$

this determinant being of order mnN.

We shall see in the next chapter how, when $m = n = 1$, the condition of Theorem 1.12 is related to the classical theory of the resultant of two polynomials.

In fact in Theorem 1.12 only one of the matrices $A(\lambda)$, $B(\lambda)$ need be regular. For example, suppose $A(\lambda)$ has leading matrix A_0 which is singular and consider instead of (1.2.17) the matrix

$$S = A_0 \otimes D^N + A_1 \otimes D^{N-1} + \cdots + A_N \otimes I_{mM}.$$

If $A(\lambda)$ has (ij) element $a_{ij}^0\lambda^N + a_{ij}^1\lambda^{N-1} + \cdots + a_{ij}^N$ then the (ij) submatrix of S is $a_{ij}^0D^N + a_{ij}^1D^{N-1} + \cdots + a_{ij}^NI_{mM}$. It therefore follows that if the latent roots of $A(\lambda)$ are $\theta_1, \theta_2, \ldots, \theta_k$ so that

$$\det A(\lambda) = g(\lambda) = c(\lambda - \theta_1)^{t_1}(\lambda - \theta_2)^{t_2} \ldots (\lambda - \theta_k)^{t_k}$$

then

$$\det S = \det g(D) = c \det(D - \theta_1 I_{mM})^{t_1} \ldots (D - \theta_k I_{mM})^{t_k}$$

(c is a constant). Thus S is singular if and only if $D - \theta_i I_{mM}$ is singular for some i, i.e. if and only if $A(\lambda)$ and $B(\lambda)$ have a common latent root (since the roots of $B(\lambda)$ are the same as the characteristic roots of D).

EXERCISES

1.2.1 [38] Prove the necessity of the condition in Theorem 1.10 by multiplying $\begin{bmatrix} A & E \\ 0 & B \end{bmatrix}$ on the left and right by suitable invertible partitioned matrices.

1.2.2 Show that the matrix $A^2 + b_1A + b_2I_n$ is singular for some $b_1, b_2 > 0$ if and only if A has at least one characteristic root with negative real part.

1.3 Degrees of greatest common divisors of invariant factors of two regular matrices

Let the matrix of (1.2.18) be denoted by

$$R = I_m \otimes C^M + B_1 \otimes C^{M-1} + \cdots + B_M \otimes I_{nN}. \quad (1.3.1)$$

Not only does R give us the necessary and sufficient condition of

Theorem 1.12, but we can also derive further information from it as follows. Let $a_1(\lambda), a_2(\lambda), \ldots, a_n(\lambda)$ and $b_1(\lambda), b_2(\lambda), \ldots, b_m(\lambda)$ be the invariant factors of $A(\lambda)$ and $B(\lambda)$ respectively, and let δ_{ij} denote the degree of the g.c.d. of $b_i(\lambda)$ and $a_j(\lambda)$. Then:

THEOREM 1.13 [5]

$$\sum_{i=1}^{m} \sum_{j=1}^{n} \delta_{ij} = mnN - \text{rank } R.$$

Proof. As in (1.1.2) we can write

$$b_i(\lambda) = (\lambda - \lambda_1)^{\alpha_{i1}}(\lambda - \lambda_2)^{\alpha_{i2}} \ldots (\lambda - \lambda_s)^{\alpha_{is}}, \quad i = 1, 2, \ldots, m \tag{1.3.2}$$

and

$$a_j(\lambda) = (\lambda - \lambda_1)^{\beta_{j1}}(\lambda - \lambda_2)^{\beta_{j2}} \ldots (\lambda - \lambda_s)^{\beta_{js}}, \quad j = 1, 2, \ldots, n \tag{1.3.3}$$

where the λ_i are the distinct latent roots of $A(\lambda)$ and $B(\lambda)$, and $\alpha_{ip}, \beta_{jp} \geqslant 0$. From (1.3.2) and (1.3.3) we have

$$\delta_{ij} = \sum_{p=1}^{s} \min(\alpha_{ip}, \beta_{jp}), \tag{1.3.4}$$

and since the degree of det $A(\lambda)$ is nN,

$$\sum_{j=1}^{n} \sum_{p=1}^{s} \beta_{jp} = nN. \tag{1.3.5}$$

Consider any set of λ-equivalence transformations which reduce $B(\lambda)$ to its Smith form

$$\text{diag}\,[b_1(\lambda), b_2(\lambda), \ldots, b_m(\lambda)].$$

Regarding R in (1.3.1) as an $m \times m$ matrix of $nN \times nN$ submatrices, the crucial step in the proof is to realize that if these same transformations are applied as C-operations to the submatrices, R is reduced to the equivalent matrix

$$\text{diag}\,[b_1(C), b_2(C), \ldots, b_m(C)],$$

so that

$$\text{rank } R = \sum_{i=1}^{m} \text{rank } b_i(C). \tag{1.3.6}$$

Next, from (1.3.2),

$$\begin{aligned} \text{rank } b_i(C) &= \text{rank}\,(C - \lambda_1 I_{nN})^{\alpha_i} \ldots (C - \lambda_s I_{nN})^{\alpha_{is}} \\ &= \text{rank}\,(J - \lambda_1 I_{nN})^{\alpha_{i1}} \ldots (J - \lambda_s I_{nN})^{\alpha_{is}} \end{aligned} \tag{1.3.7}$$

where J is the Jordan form of C.

The elementary divisors of C are in fact just those of $A(\lambda)$; for in $\lambda I_{nN} - C$ all $n \times n$ block matrices commute with each other, so by an argument analogous to that which may be used in the scalar case to show that a companion form matrix is nonderogatory (see Exercise 1.1.8) it follows that $\lambda I_{nN} - C$ is equivalent to

$$\text{diag}\,[I_{nN}, I_{nN}, \ldots, I_{nN}, A(\lambda)].$$

Thus the Jordan form of C may be written down from (1.3.3) since the elementary divisors of $A(\lambda)$ are the factors of all the a_j (see Theorem 1.6). We have

$$\mathcal{J} = \text{diag}\,[\mathcal{J}_{\beta_{11}}(\lambda_1), \mathcal{J}_{\beta_{21}}(\lambda_1), \ldots, \mathcal{J}_{\beta_{n1}}(\lambda_1), \mathcal{J}_{\beta_{12}}(\lambda_2), \ldots, \mathcal{J}_{\beta_{ns}}(\lambda_s)],$$

where $\mathcal{J}_k(\lambda)$ is a Jordan block of order k, so

$$\mathcal{J} - \lambda_p I_{nN} = \text{diag}\,[\mathcal{J}_{\beta_{11}}(\lambda_1 - \lambda_p), \ldots, \mathcal{J}_{\beta_{1p}}(0), \ldots, \mathcal{J}_{\beta_{np}}(0), \ldots, \mathcal{J}_{\beta_{ns}}(\lambda_s - \lambda_p)]. \quad (1.3.8)$$

Since the λ_i are distinct and since rank $\{\mathcal{J}_k(0)\}^h$ is zero if $h \geqslant k$ and $k - h$ if $h < k$, substitution of (1.3.8) into (1.3.7) leads to

$$\text{rank}\, b_i(C) = \sum_j \sum_p (\beta_{jp} - \alpha_{ip}), \qquad (1.3.9)$$

the sum in (1.3.9) being taken over all j, p such that $0 < \alpha_{ip} < \beta_{jp}$. Thus from (1.3.6) and (1.3.9) we have

$$\text{rank}\, R = \sum_{i=1}^{m} \sum\sum_{\alpha_{ip} < \beta_{jp}} (\beta_{jp} - \alpha_{ip}). \qquad (1.3.10)$$

The theorem now follows by writing (1.3.4) as

$$\delta_{ij} = \sum_{\alpha_{ip} \geqslant \beta_{jp}} \beta_{jp} + \sum_{\alpha_{ip} < \beta_{jp}} \alpha_{ip}$$

so that

$$\sum_{i=1}^{m}\sum_{j=1}^{n} \delta_{ij} = \sum\sum\sum_{\alpha_{ip} \geqslant \beta_{jp}} \beta_{jp} + \sum\sum\sum_{\alpha_{ip} < \beta_{jp}} \alpha_{ip} \qquad (1.3.11)$$

and addition of (1.3.10) and (1.3.11) gives

$$\begin{aligned} \text{rank}\, R + \sum_{i=1}^{m}\sum_{j=1}^{n} \delta_{ij} &= \sum_{i=1}^{m}\sum_{j=1}^{n}\sum_{p=1}^{s} \beta_{jp} \\ &= \sum_{i=1}^{m} nN \qquad \text{by (1.3.5)} \\ &= mnN. \end{aligned}$$

The reader should satisfy himself that the preceding proof carries over even if $B(\lambda)$ is rectangular provided $b_m \not\equiv 0$, m now being the smaller of the dimensions of $B(\lambda)$.

By using the property that $a_i(\lambda)$ divides $a_{i+1}(\lambda)$ and $b_j(\lambda)$ divides $b_{j+1}(\lambda)$ (Theorem 1.5) information about relatively prime pairs of invariant factors can be obtained:

THEOREM 1.14 If

$$\text{rank } R > \max \{mnN - n + k - 2, mnN - m + k - 2, \\ mnN - (n - k + 1)(m - x + 1)\}$$

then $\delta_{ij} = 0$ for $i \leqslant k - 1, j = 1, 2, \ldots, n; j \leqslant k - 1, i = 1, 2, \ldots, m$; and $i = 1, 2, \ldots, x, j = k$.

The proof of this theorem is straightforward but tedious and the reader is therefore referred to [5] for details.

Clearly, by interchanging $A(\lambda)$ and $B(\lambda)$, equivalent theorems may be easily formulated in terms of the mnM order matrix

$$I_n \otimes D^N + A_1 \otimes D^{N-1} + \cdots + A_N \otimes I_{mM}.$$

EXERCISES

1.3.1 Obtain an expression for the degree of the greatest common divisor of two scalar polynomials by specializing Theorem 1.13.

1.3.2 It can be shown that the number of linearly independent solutions of $AX = XB$ (A and B are given constant $n \times n$ matrices) is $\sum_i \sum_j e_{ij}$, where e_{ij} is the degree of the g.c.d. of the ith invariant factor of $\lambda I_n - A$ and the jth invariant factor of $\lambda I_n - B$ [22, p. 90]. Use Theorem 1.13 to show that this number of solutions is $n^2 - \text{rank}\,(I_n \otimes A - B \otimes I_n)$.

1.4 Linear system theory; relatively prime matrices

A large part of modern control theory is concerned with the analysis and design of linear multivariable systems. Many methods have been suggested [23] but only the approach developed by Rosenbrock [24–37] will be detailed here since it is intimately related to the theory of polynomial matrices. Readers interested in a formulation of the problem based on abstract algebra should consult the article by Kalman in a recent book [20], and a basic reference on linear control systems is [44].

A constant linear dynamical system can be represented by a set of n first order linear differential equations

$$\dot{x} = Ax + Bu \tag{1.4.1}$$

where A $(n \times n)$ and B $(n \times l)$ are matrices with complex elements, $x = [x_1(t), x_2(t), \ldots, x_n(t)]^T$ is a vector representing the 'state' of the system and $u = [u_1(t), u_2(t), \ldots, u_l(t)]^T$ is the vector of 'inputs' or controlled variables. After Laplace transformation, assuming $x(0) = 0$, equation (1.4.1) becomes

$$(\lambda I_n - A)\bar{x} = B\bar{u}. \tag{1.4.2}$$

[$(\bar{\ })$ denotes Laplace transform; in keeping with the rest of this chapter we use λ rather than the familiar s.]

If $y = [y_1(t), \ldots, y_m(t)]^T$ is the vector of measured variables or 'outputs' we have in addition

$$\bar{y} = C\bar{x} + D(\lambda)\bar{u}, \tag{1.4.3}$$

C $(m \times n)$ being a matrix with complex elements and $D(\lambda)$ $(m \times l)$ a polynomial matrix. (In fact in many practical situations $D(\lambda)$ turns out to be either zero or a constant matrix.) Solving for $\bar{y}$ between (1.4.2) and (1.4.3) gives

$$\bar{y} = G\bar{u}$$

where

$$G = C(\lambda I_n - A)^{-1}B + D(\lambda) \tag{1.4.4}$$

is the so-called *transfer function matrix*. Notice that the elements of G are *rational* functions of λ (see Chapter 3). The properties of the system described by (1.4.1) and (1.4.3) may be concentrated in the $(m + n) \times (l + n)$ *first form* (or *state space form*) *system matrix*

$$P(\lambda) = \begin{bmatrix} \lambda I_n - A & B \\ -C & D(\lambda) \end{bmatrix}. \tag{1.4.5}$$

A more general form of the system equations can be written, after Laplace transformation, as [26]

$$\left.\begin{aligned} T(\lambda)\bar{z} &= U(\lambda)\bar{u} \\ \bar{y} &= V(\lambda)\bar{z} + W(\lambda)\bar{u} \end{aligned}\right\} \tag{1.4.6}$$

where $T(\lambda)$, $U(\lambda)$, $V(\lambda)$ and $W(\lambda)$ are polynomial matrices with dimensions $r \times r$, $r \times l$, $m \times r$ and $m \times l$ respectively, and $\det T(\lambda) \not\equiv 0$. The transfer function matrix becomes

$$G = VT^{-1}U + W \tag{1.4.7}$$

and the *second form system matrix* is

$$P(\lambda) = \begin{bmatrix} T(\lambda) & U(\lambda) \\ -V(\lambda) & W(\lambda) \end{bmatrix}. \tag{1.4.8}$$

The *order* of the system represented by (1.4.6) is the degree n of $\det T(\lambda)$. It can be assumed for convenience and without loss of generality that $r \geqslant n$ [35]. Two different sets of matrices $T(\lambda)$, $U(\lambda)$, $V(\lambda)$, $W(\lambda)$ can give rise to the same G (see Exercise 1.4.1), and an important question is whether a system matrix $P(\lambda)$ in second form has least order. To answer this we need the following definition [32]: The matrices $T(\lambda)$ and $U(\lambda)$ are said to be *relatively prime* if the Smith form of $[T(\lambda) \ \ U(\lambda)]$ is $[I_r \ \ 0]$. Clearly when $r = l = 1$ this reduces to the usual definition of relative primeness of two polynomials. In fact both Smith [41] (see Chapter 7, Section 1) and Châtelet† [8] have used the term in a similar sense in connection with matrices having integer elements. Notice that when $l = r$, relative primeness of $\det T(\lambda)$ and $\det U(\lambda)$ is only a sufficient condition for $T(\lambda)$ and $U(\lambda)$ to be relatively

† *See also* [22], *pages* 36, 44, *for details of Châtelet's results.*

prime. It should also be realized that a knowledge of the invariant factors of $T(\lambda)$ and $U(\lambda)$ is not sufficient to determine whether $T(\lambda)$ and $U(\lambda)$ are relatively prime, as the following very simple example illustrates:

$$T(\lambda) = \begin{bmatrix} 1 & 0 \\ 0 & \lambda \end{bmatrix}, \; U(\lambda) = \begin{bmatrix} 1 & 0 \\ 0 & \lambda \end{bmatrix}, \; [T(\lambda) \quad U(\lambda)] \sim \begin{bmatrix} 1 & 0 & 0 & 0 \\ 0 & \lambda & 0 & 0 \end{bmatrix},$$

$$T(\lambda) = \begin{bmatrix} 1 & 0 \\ 0 & \lambda \end{bmatrix}, \; U(\lambda) = \begin{bmatrix} \lambda & 0 \\ 0 & 1 \end{bmatrix}, \; [T(\lambda) \quad U(\lambda)] \sim \begin{bmatrix} 1 & 0 & 0 & 0 \\ 0 & 1 & 0 & 0 \end{bmatrix}.$$

In both cases $T(\lambda)$ and $U(\lambda)$ have invariant factors $1, \lambda$. Thus the results of the previous section on invariant factors unfortunately do not seem to have application here.

The importance for linear system theory of the above definition is made clear by:

THEOREM 1.15 [26] The matrix $P(\lambda)$ in (1.4.8) has least order if and only if $T(\lambda)$ and $U(\lambda)$ are relatively prime and $T^T(\lambda)$ and $V^T(\lambda)$ are relatively prime.

We now give some interesting results concerning relatively prime polynomial matrices.

THEOREM 1.16 [32] If $T(\lambda)$ and $U(\lambda)$ are relatively prime and $l = m$ then there exist polynomial matrices $V(\lambda)$, $W(\lambda)$ such that $P(\lambda)$ in (1.4.8) has Smith form diag $[I_r, \Phi(\lambda)]$ where $\Phi(\lambda)$ is any $m \times m$ non-singular diagonal matrix satisfying the conditions of Theorem 1.5.

Proof. Since $[T(\lambda) \; U(\lambda)]$ has Smith form $[I_r \; 0]$ there exist, by virtue of Theorems 1.3 and 1.5, invertible polynomial matrices $X(\lambda)$, $Y(\lambda)$ of orders r and $(r + l)$ respectively such that

$$\begin{aligned} [T(\lambda) \quad U(\lambda)] &= X(\lambda)\,[I_r \quad 0]\; Y(\lambda) \qquad (1.4.9) \\ &= X(\lambda)\,[I_r \quad 0] \begin{bmatrix} Y_1(\lambda) & Y_2(\lambda) \\ Y_3(\lambda) & Y_4(\lambda) \end{bmatrix} \begin{matrix} r \\ l \end{matrix} \\ &\qquad\qquad\qquad\quad\; \begin{matrix} r \qquad\quad & l \end{matrix} \\ &= X(\lambda)\,[Y_1(\lambda) \quad Y_2(\lambda)] \end{aligned}$$

so

$$T(\lambda) = X(\lambda) Y_1(\lambda), \; U(\lambda) = X(\lambda) Y_2(\lambda). \qquad (1.4.10)$$

Now

$$\begin{bmatrix} X & 0 \\ 0 & I_m \end{bmatrix} \begin{bmatrix} I_r & 0 \\ 0 & \Phi \end{bmatrix} \begin{bmatrix} Y_1 & Y_2 \\ Y_3 & Y_4 \end{bmatrix} = \begin{bmatrix} XY_1 & XY_2 \\ \Phi Y_3 & \Phi Y_4 \end{bmatrix} = \begin{bmatrix} T & U \\ \Phi Y_3 & \Phi Y_4 \end{bmatrix}$$

by (1.4.10). Hence taking $V = -\Phi Y_3$, $W = \Phi Y_4$ gives $P(\lambda)$ the desired Smith form.

THEOREM 1.17 [26] A necessary and sufficient condition that $T(\lambda)$ and $U(\lambda)$ be relatively prime is that the matrix $[T(\lambda_i) \; U(\lambda_i)]$ has rank r for each value of $i = 1, 2, \ldots, n$, where λ_i are the latent roots of $T(\lambda)$.

Proof. If $[T(\lambda) \; U(\lambda)]$ has Smith form $[I_r \; 0]$ this implies that the rth determinantal divisor of $[T(\lambda) \; U(\lambda)]$ is unity, so $[T(\lambda) \; U(\lambda)]$ has rank r for all possible values of λ.

Conversely, if $[T(\lambda) \; U(\lambda)]$ has Smith form $[S(\lambda) \; 0]$ where $S(\lambda) = \text{diag}\,[s_1(\lambda), s_2(\lambda), \ldots, s_r(\lambda)]$ then if $s_r(\mu) = 0$ this implies that μ is a zero of the rth determinantal divisor of $[T(\lambda) \; U(\lambda)]$ and hence that μ is a latent root of $T(\lambda)$. Further, the rank of $[S(\lambda) \; 0]$ is the same as the rank of $[T(\lambda) \; U(\lambda)]$ for all values of λ. Since this is r (by assumption) for all possible μ, it follows that $s_r(\lambda) = 1$ and hence that $S(\lambda) = I_r$.

THEOREM 1.18 [35] A necessary and sufficient condition for $T(\lambda)$ and $U(\lambda)$ to be relatively prime is that there exist polynomial matrices $Q(\lambda)$ $(r \times r)$ and $N(\lambda)$ $(l \times r)$ such that

$$T(\lambda)Q(\lambda) + U(\lambda)N(\lambda) = I_r. \tag{1.4.11}$$

Proof. If $T(\lambda)$ and $U(\lambda)$ are relatively prime then from (1.4.9) we have

$$[T(\lambda) \quad U(\lambda)]Z(\lambda) = X(\lambda)[I_r \quad 0],$$

where $Z(\lambda) = Y^{-1}(\lambda)$ is also a polynomial matrix. Thus we can write

$$[T(\lambda) \quad U(\lambda)] \begin{bmatrix} Z_1(\lambda) & Z_2(\lambda) \\ Z_3(\lambda) & Z_4(\lambda) \end{bmatrix} = X(\lambda)[I_r \quad 0]$$

whence

$$T(\lambda)Z_1(\lambda) + U(\lambda)Z_3(\lambda) = X(\lambda)$$

or

$$T(\lambda)[Z_1(\lambda)X^{-1}(\lambda)] + U(\lambda)[Z_3(\lambda)X^{-1}(\lambda)] = I_r.$$

Conversely, if

$$[T(\lambda) \quad U(\lambda)] \begin{bmatrix} Q(\lambda) \\ N(\lambda) \end{bmatrix} = I_r$$

it follows that the rank of $[T(\lambda) \; U(\lambda)]$ must be r for all possible values of λ, and hence that $T(\lambda)$, $U(\lambda)$ are relatively prime by Theorem 1.17.

Theorem 1.18 is closely related to the following ideas (see [22], p. 35): if $D_i(\lambda)$ is a matrix such that $T(\lambda) = D_i(\lambda)T_i(\lambda)$, $U(\lambda) = D_i(\lambda)U_i(\lambda)$, where $T_i(\lambda)$ and $U_i(\lambda)$ are also polynomial matrices, then $D_i(\lambda)$ is called a *common left divisor* of $T(\lambda)$ and $U(\lambda)$. If $\tilde{D}(\lambda)$ is a common left divisor such that $\tilde{D}(\lambda) = D_i(\lambda)X_i(\lambda)$ for all such matrices $D_i(\lambda)$ then $\tilde{D}(\lambda)$ is called a *greatest* common left divisor of $T(\lambda)$, $U(\lambda)$. We then have an alternative to Theorem 1.18:

THEOREM 1.19 $T(\lambda)$ and $U(\lambda)$ are relatively prime if and only if $\tilde{D}(\lambda)$ is invertible.

Proof. If $T(\lambda)$ and $U(\lambda)$ are relatively prime then from Theorem 1.18 we have

$$T(\lambda)Q(\lambda) + U(\lambda)N(\lambda) = I_r.$$

Since $T(\lambda) = \tilde{D}(\lambda)\tilde{T}(\lambda)$, $U(\lambda) = \tilde{D}(\lambda)\tilde{U}(\lambda)$, this gives

$$\tilde{D}(\lambda)\{\tilde{T}(\lambda)Q(\lambda) + \tilde{U}(\lambda)N(\lambda)\} = I_r$$

showing that $\tilde{D}(\lambda)$ is invertible.

Conversely, it is easy to show (see Exercise 1.4.5) that there exist matrices $L(\lambda)$, $M(\lambda)$ such that

$$T(\lambda)L(\lambda) + U(\lambda)M(\lambda) = \tilde{D}(\lambda).$$

If $\tilde{D}(\lambda)$ is invertible,

$$T(\lambda)L(\lambda)\tilde{D}^{-1}(\lambda) + U(\lambda)M(\lambda)\tilde{D}^{-1}(\lambda) = I_r.$$

Hence, by Theorem 1.18, $T(\lambda)$ and $U(\lambda)$ are relatively prime.

Again we see that when $l = r = 1$, Theorem 1.19 reduces to the usual result for polynomials.

The following example illustrates the relationship between Theorems 1.15 and 1.19. If

$$T(\lambda) = \begin{bmatrix} 1 & \lambda + 1 \\ \lambda + 1 & \lambda + 3 \end{bmatrix} \quad \text{and} \quad U(\lambda) = \begin{bmatrix} \lambda + 2 \\ 6 \end{bmatrix}$$

then these matrices are not relatively prime since

$$[T(\lambda)\ U(\lambda)] \sim \begin{bmatrix} 1 & 0 & 0 \\ 0 & \lambda - 1 & 0 \end{bmatrix}.$$

In $T^{-1}(\lambda)U(\lambda)$ it is easily verified that a common factor $(\lambda - 1)$ can be cancelled out. In fact

$$T(\lambda) = \begin{bmatrix} 1 & 0 \\ 2 & \lambda - 1 \end{bmatrix} \begin{bmatrix} 1 & \lambda + 1 \\ 1 & -1 \end{bmatrix}, \; U(\lambda) = \begin{bmatrix} 1 & 0 \\ 2 & \lambda - 1 \end{bmatrix} \begin{bmatrix} \lambda + 2 \\ -2 \end{bmatrix}$$

so

$$T^{-1}(\lambda)U(\lambda) = \begin{bmatrix} 1 & \lambda + 1 \\ 1 & -1 \end{bmatrix}^{-1} \begin{bmatrix} \lambda + 2 \\ -2 \end{bmatrix}.$$

The reader is also advised at this point to attempt Exercise 1.4.6.

It is interesting to notice that we now have three quite different generalizations of the basic theorem on polynomials quoted at the beginning of Section 1.2. The first, for two square polynomial matrices, was provided by Theorem 1.8, and the second, for the $r \times r$ and $r \times l$ matrices $T(\lambda)$ and $U(\lambda)$, by Theorem 1.18. It is easy to see that Theorem 1.16 also provides a generalization if we restate the theorem on polynomials as follows: if $t(\lambda)$ and $u(\lambda)$ are relatively prime then for an arbitrary polynomial $\phi(\lambda)$ there exist polynomials $v(\lambda)$, $w(\lambda)$ such that $tw + uv = \phi$, or equivalently, such that

$$\begin{vmatrix} t(\lambda) & u(\lambda) \\ -v(\lambda) & w(\lambda) \end{vmatrix} = \phi(\lambda).$$

Since $t(\lambda)$ and $u(\lambda)$ are relatively prime this implies that the Smith form of

$$\begin{bmatrix} t(\lambda) & u(\lambda) \\ -v(\lambda) & w(\lambda) \end{bmatrix} \text{ is } \begin{bmatrix} 1 & 0 \\ 0 & \phi(\lambda) \end{bmatrix},$$

and the relationship with Theorem 1.16 is now clear.

In Theorem 1.8 it was possible to give information concerning the degrees of the matrices involved, as in the case of polynomials. However, no similar results are known for Theorems 1.16 or 1.18. The following examples of possible cases of equation (1.4.11) serve to illustrate some of the difficulties involved.

(a) $t(\lambda)q(\lambda) + [u_1(\lambda) \quad u_2(\lambda)]\begin{bmatrix} n_1(\lambda) \\ n_2(\lambda) \end{bmatrix} = 1.$

If $u_1(\lambda)$ and $u_2(\lambda)$ are relatively prime then we can take $q(\lambda) = 0$.

(b) $\begin{bmatrix} 1 & t_1(\lambda) \\ 0 & t_2(\lambda) \end{bmatrix}\begin{bmatrix} 1 & -u(\lambda) \\ 0 & 0 \end{bmatrix} + \begin{bmatrix} u(\lambda) \\ 1 \end{bmatrix}[0 \quad 1] = I_2.$

In both cases $\delta\{Q(\lambda)\}$ and $\delta\{N(\lambda)\}$ are independent of $\delta\{T(\lambda)\}$ and also of $\delta\{\det T(\lambda)\}$.

However, when $T(\lambda) = \lambda I_n - A$ and $U(\lambda) = B$ [as in (1.4.5)] it can be shown that the degrees of $Q(\lambda)$ and $N(\lambda)$ are dependent on the degree of $\det T(\lambda)$. More precisely,

THEOREM 1.20 [34, 37] Either of the following is a necessary and sufficient condition for the matrices $\lambda I_n - A$ and B ($n \times l$) to be relatively prime.

(i) Given any polynomial n-vector $z(\lambda)$ with $\delta\{z(\lambda)\} \leqslant n - 1$ there exist a polynomial n-vector $x(\lambda)$ and a polynomial l-vector $y(\lambda)$ with $\delta\{x(\lambda)\} \leqslant n - 2$, $\delta\{y(\lambda)\} \leqslant n - 1$ such that

$$(\lambda I_n - A)x(\lambda) + By(\lambda) = z(\lambda). \tag{1.4.12}$$

(ii) There exist polynomial matrices $X(\lambda)$ ($n \times n$) and $Y(\lambda)$ ($l \times n$) with $\delta\{X(\lambda)\} \leqslant n - r - 1$, $\delta\{Y(\lambda)\} \leqslant n - r$, where $r = \text{rank } B$, such that

$$(\lambda I_n - A)X(\lambda) + BY(\lambda) = I_n. \tag{1.4.13}$$

Two results closely relating to the preceding one are now given.

THEOREM 1.21 [34] The matrices $\lambda I_n - A$ and B are relatively prime if and only if the matrix

$$\underbrace{\begin{bmatrix} I_n & 0 & & & & & & & B \\ -A & I_n & & & & & & \cdot & \\ 0 & -A & & & & & \cdot & & \\ & & \cdot & & & \cdot & & & \\ & & & \cdot & \cdot & & & & \\ & & & & I_n \; 0 & B & & & \\ & & & & -A \; B & 0 & & & \end{bmatrix}}_{\leftarrow (n-1) \text{ blocks} \rightarrow \leftarrow n \text{ blocks} \rightarrow} \updownarrow n \text{ blocks}$$

has rank n^2.

It is interesting to compare this with (1.2.15); see also [33] for yet another generalization of resultant.

THEOREM 1.22 [26] The matrices $\lambda I_n - A$ and B are relatively prime if and only if the matrix $[B, AB, A^2B, \ldots, A^{n-1}B]$ has rank n.

The condition of Theorem 1.22 is in fact the well-known controllability criterion for a system in the form (1.4.1) (see Appendix 2).

We have only given here a selection of results from Rosenbrock's many papers on linear systems theory, choosing some which we feel to be of special interest in relation to polynomial matrices. Further results are given in Chapter 3, but for full details we must refer the reader to Rosenbrock's original papers [24–35] and to his text [36] which gives a complete account. For a partial solution of a particular problem arising in Rosenbrock's work see [2], and also of interest are two papers by Kalman [18, 19].

EXERCISES

1.4.1 [25] Show that if $P_1(\lambda)$ and $P_2(\lambda)$ are two system matrices in the form (1.4.8) and are related by

$$P_2(\lambda) = \begin{bmatrix} M_1(\lambda) & 0 \\ N_1(\lambda) & I_m \end{bmatrix} P_1(\lambda) \begin{bmatrix} M_2(\lambda) & N_2(\lambda) \\ 0 & I_l \end{bmatrix},$$

where $M_1(\lambda)$ and $M_2(\lambda)$ are invertible matrices of order r and $N_1(\lambda)$ and $N_2(\lambda)$ are $m \times r$ and $r \times l$ respectively, then the transfer function matrices are the same.

1.4.2 [34] In equation (1.4.12) write $x(\lambda) = x_0 + x_1\lambda + \cdots + x_{n-2}\lambda^{n-2}, y(\lambda) = y_0 + \cdots + y_{n-1}\lambda^{n-1}$ and $z(\lambda) = z_0 + \cdots + z_{n-1}\lambda^{n-1}$, equate powers of λ and hence verify the necessity of condition (i) of Theorem 1.20. By taking

$x^{(i)}(\lambda)$ and $y^{(i)}(\lambda)$ to be the particular solution of (1.4.12) when $z^{(i)}(\lambda)$ is the ith column e_i of I_n, show also that

$$X = [x^{(1)}(\lambda), \ldots, x^{(n)}(\lambda)], Y = [y^{(1)}(\lambda), \ldots, y^{(n)}(\lambda)]$$

satisfy (1.4.13).

1.4.3 [34] Let (1.4.13) have a solution and show that

$$[A^{n-1}B, A^{n-2}B, \ldots, B]\begin{bmatrix} y_{n-1} \\ y_{n-2} \\ \cdot \\ \cdot \\ \cdot \\ y_0 \end{bmatrix} = e_i \qquad (i = 1, 2, \ldots, n).$$

Hence establish the sufficiency of (ii) in Theorem 1.20.

1.4.4 If $B(\lambda)$ is of order n and degree k, show that if $B(\lambda)$ and $\lambda I_n - A$ are relatively prime then $A^kB_0 + A^{k-1}B_1 + \cdots + B_k \neq 0$ (see Exercise 1.1.1).

1.4.5 Using an argument along the lines of the proof of the first half of Theorem 1.18, show that if $D(\lambda)$ is a greatest common left divisor of $T(\lambda)$ and $U(\lambda)$ then there exist matrices $L(\lambda)$, $M(\lambda)$ such that $T(\lambda)L(\lambda) + U(\lambda)M(\lambda) = D(\lambda)$.

1.4.6 Let $D(\lambda)$ be as in the previous exercise, with $T(\lambda) = D(\lambda)T_1(\lambda)$, $U(\lambda) = D(\lambda)U_1(\lambda)$ and let $E(\lambda)$ be a greatest common right divisor of $T(\lambda)$ and $V(\lambda)$ with $T(\lambda) = T_2(\lambda)E(\lambda)$, $V(\lambda) = V_2(\lambda)E(\lambda)$. Consider the systems with matrices $T_1(\lambda)$, $U_1(\lambda)$, $V(\lambda)$, $W(\lambda)$ and $T_2(\lambda)$, $U(\lambda)$, $V_2(\lambda)$, $W(\lambda)$. Show that they both have the same transfer function matrix (1.4.7) and that they have order less than that for (1.4.8) unless $D(\lambda)$ and $E(\lambda)$ are invertible.

1.5 A simple class of parametric linear programming problems

We conclude this chapter with an application of polynomial matrices to linear programming. The standard linear programming problem may be written:

Choose the column n-vector x so as to minimize

$$z = cx \tag{1.5.1}$$

subject to

$$x \geqslant 0 \tag{1.5.2}$$

and

$$\alpha x = b, \tag{1.5.3}$$

where α is an $m \times n$ matrix with constant (usually real) elements, b a column m-vector and c a row n-vector (again usually having real components). Suppose that an optimal solution has been found so that (1.5.3) may be written without loss of generality as

$$Ax = b - Ny, \tag{1.5.4}$$

where A is a nonsingular $m \times m$ matrix, N an $m \times (n - m)$ matrix, $x = [x_1, x_2, \ldots, x_m]^T$ is now the vector of optimal basic variables and $y = [x_{m+1}, \ldots, x_n]^T$ is the vector of non-basic variables. From (1.5.4) we have

$$x = A^{-1}b - A^{-1}Ny, \tag{1.5.5}$$

and since $y = 0$ gives an optimal feasible solution to the original problem we can write the optimal vector as

$$x^0 = [x_1^0, x_2^0, \ldots, x_m^0]^T = A^{-1}b$$

and $x_i^0 \geqslant 0$, all i (from (1.5.2)). Substitution of (1.5.5) into (1.5.1) gives

$$\begin{aligned} z &= c_x A^{-1}b + (c_y - c_x A^{-1}N)y \\ &= c_x A^{-1}b + \sum_{i=m+1}^{n} \gamma_i x_i \end{aligned} \tag{1.5.6}$$

where $c_x = [c_1, c_2, \ldots, c_m]$, $c_y = [c_{m+1}, \ldots, c_n]$ and each $\gamma_i \geqslant 0$. For further details of basic linear programming theory we refer the reader to a standard text, such as [11] or [15].

Since the coefficients in problems arising from practical situations are often not known exactly, a considerable amount of attention has been paid to analysis of the sensitivity of the solution of linear programming problems to variations in the coefficients. Chapter 8 of [15] and papers by Shetty [40] and Courtillot [9] deal with variations in the elements of α, but generally the methods and algorithms involved are quite complicated. Recently Willner [43] has studied the problem where α in (1.5.3) is replaced by $\alpha + \lambda\beta$, λ being a parameter and β an $m \times n$ constant matrix. He emphasizes the difficulties which are encountered but gives an algorithm for the case when β is of rank one. Williams [42] gives theorems on the existence of solutions when α, b and c all vary parametrically and $\lambda \rightarrow 0$.

However, an approach which has had some success is to find changes which can occur in the coefficients without affecting the choice of optimal feasible basic variables. This procedure was first developed by Saaty and Gass [39] for changes in b or c, and was also used by the author [1] to deal with changes in a single column of α.

We now develop this latter technique further and construct a class of parametric problems that admits a very simple solution. The situation studied is to find the region within which the parameter λ may lie so that when A in (1.5.4) changes to $A + \lambda B$ (B being an $m \times m$ constant matrix) the set of optimal basic variables is unaltered. The novelty of the approach is to allow only matrices B such that $A + \lambda B$ is invertible and additionally has an inverse which is linear in λ. We can then obtain

THEOREM 1.23 [3] Let A in (1.5.4) be replaced by $A + \lambda B$ where $B = AC$ or CA, $C = TDT^{-1}$ (T is an arbitrary nonsingular $m \times m$ matrix) and

$$D = \text{diag}\left[0, 0, \ldots, \begin{bmatrix} 0 & 1 \\ 0 & 0 \end{bmatrix}, \ldots\right], \tag{1.5.7}$$

the number and sequence of the diagonal entries of each type in (1.5.7) being entirely arbitrary. Then $x_1, x_2, \ldots, x_m$ remain the optimal basic variables provided

$$\max(\underline{\mu}, \underline{\theta}) \leqslant \lambda \leqslant \min(\bar{\mu}, \bar{\theta}) \tag{1.5.8}$$

where

$$\left.\begin{aligned} \bar{\theta} &= \min_i (x_i/\theta_i) && \text{for } \theta_i > 0 \\ &= \infty && \text{if no } \theta_i > 0 \end{aligned}\right\} \tag{1.5.9}$$

$$\left.\begin{aligned} \underline{\theta} &= \max_i (x_i/\theta_i) && \text{for } \theta_i < 0 \\ &= -\infty && \text{if no } \theta_i < 0 \end{aligned}\right\} \tag{1.5.10}$$

$$\left.\begin{aligned} \bar{\mu} &= \min_i (-\gamma_i/\mu_i) && \text{for } \mu_i < 0 \\ &= \infty && \text{if no } \mu_i < 0 \end{aligned}\right\} \tag{1.5.11}$$

$$\left.\begin{aligned} \underline{\mu} &= \max_i (-\gamma_i/\mu_i) && \text{for } \mu_i > 0 \\ &= -\infty && \text{if no } \mu_i > 0 \end{aligned}\right\} \tag{1.5.12}$$

and $A^{-1}Bx^0 = [\theta_1, \theta_2, \ldots, \theta_m]^T = \theta$, $c_x A^{-1}BA^{-1}N = [\mu_{m+1}, \ldots, \mu_n]$. Further, the solution to the parametric problem is then $x^0 - \lambda\theta$, and the associated minimum value of z is $z^0 - \lambda c_x \theta$.

Proof. First, it is easy to show that

$$(A + \lambda B)^{-1} = A^{-1} - \lambda A^{-1}BA^{-1}, \tag{1.5.13}$$

provided B satisfies the condition

$$BA^{-1}B = 0. \tag{1.5.14}$$

The general solution of (1.5.14) is

$$B = AC, \tag{1.5.15a}$$

or, alternatively,

$$B = CA, \tag{1.5.15b}$$

where C is given by

$$C^2 = 0. \tag{1.5.16}$$

It is well known (e.g. [14], p. 226) that the general solution of (1.5.16) is that given in the statement of the theorem. Notice that the rank of D, and hence of B, is half the number of diagonal entries of the second type that are present in (1.5.7) and so is at most $m/2$.

Next, if (1.5.13) is used with B given by (1.5.15a) or (1.5.15b), the solution of (1.5.4) becomes

$$x = A^{-1}(I - \lambda BA^{-1})b - A^{-1}(I - \lambda BA^{-1})Ny. \qquad (1.5.17)$$

Thus, for $x_1, \ldots, x_m$ to remain a feasible basic set (i.e. $x \geqslant 0$) it follows from (1.5.17) that $A^{-1}b - \lambda A^{-1}BA^{-1}b \geqslant 0$, which reduces to

$$x^0 - \lambda A^{-1}Bx^0 \geqslant 0 \quad \text{or} \quad x_i^0 - \lambda\theta_i \geqslant 0, \quad i = 1, \ldots, m. \qquad (1.5.18)$$

By virtue of (1.5.9) and (1.5.10), the range of values of λ determined by the inequalities (1.5.18) is

$$\underline{\theta} \leqslant \lambda \leqslant \bar{\theta}. \qquad (1.5.19)$$

Clearly $\bar{\theta} \geqslant 0$, $\underline{\theta} \leqslant 0$ so that $\lambda = 0$ is within the region (1.5.19), as, of course, is already known.

Finally, the condition for the basis to remain optimal is obtained in a similar way. From (1.5.1), (1.5.6) and (1.5.17) we obtain

$$z = c_x A^{-1}(I - \lambda BA^{-1})b + [c_y - c_x A^{-1}(I - \lambda BA^{-1})N]y \qquad (1.5.20)$$

$$= c_x A^{-1}(I - \lambda BA^{-1})b + \sum_{i=m+1}^{n} (\gamma_i + \lambda\mu_i)x_i. \qquad (1.5.21)$$

Thus, for optimality (1.5.21) shows that

$$\gamma_i + \lambda\mu_i \geqslant 0 \qquad (i = m + 1, \ldots, n) \qquad (1.5.22)$$

and from (1.5.11) and (1.5.12) it follows that the inequalities (1.5.22) imply

$$\underline{\mu} \leqslant \lambda \leqslant \bar{\mu}.$$

Therefore, for the optimal feasible basic variables to be the same for the parametric problem as for the original problem, λ must lie in the range (1.5.8), and this includes $\lambda = 0$.

From (1.5.17) we see that the solution to the parametric problem is $x^0 - \lambda\theta$, and from (1.5.21) that the associated minimum value of z is $z^0 - \lambda c_x\theta$, where z^0 is the value for the original problem.

The theorem thus provides a method for a class of parametric linear programming problems that avoids in a remarkable fashion the complexity of the general case. However, nothing can be inferred for values of λ outside the region (1.5.8), for the basis will then change to some matrix A_1, say, and N to N_1, both matrices being linear in λ. Thus even if A_1 were still invertible with inverse of degree one (which is unlikely) the expression corresponding to (1.5.20) would contain terms of degree two and the range of values of λ would have to be determined from the resulting quadratic inequalities. Although Willner's algorithm [43] can be used when the choice of basic variables alters, nevertheless it would seem that the simplicity of the restricted method presented in this section makes it a potentially useful one. Since C can be chosen similar to any matrix of Jordan form (1.5.7) with predetermined rank $\leqslant m/2$, the results hold for a large class of matrices B, and thus in this respect are certainly more general than those of Willner, where the matrix β can have only unit rank.

EXERCISES

1.5.1 [3] An optimal feasible solution to the problem:

$$\text{minimize } x_2 - 3x_3 + 2x_5 \text{ subject to (1.5.2) and}$$
$$\begin{aligned} x_1 + 3x_2 - x_3 \qquad\qquad + 2x_5 \qquad &= 7 \\ - 2x_2 + 4x_3 + x_4 \qquad\qquad\qquad &= 12 \\ - 4x_2 + 3x_3 \qquad\qquad + 8x_5 + x_6 &= 10 \end{aligned}$$

is $x_2 = 4$, $x_3 = 5$, $x_6 = 11$, $x_1 = x_4 = x_5 = 0$ and $\gamma_1 = 1/5$, $\gamma_4 = 4/5$, $\gamma_5 = 12/5$, $z^0 = -11$.

Take, in Theorem 1.23,

$$D = \begin{bmatrix} 0 & 1 & 0 \\ 0 & 0 & 0 \\ 0 & 0 & 0 \end{bmatrix}, \quad T = \begin{bmatrix} 1 & 0 & 0 \\ 2 & -1 & 0 \\ 3 & 0 & -2 \end{bmatrix}$$

(the matrix T has been chosen so that the last column of B in (1.5.15a) is zero) and show that in this case x_2, x_3, x_6 remain optimal basic feasible variables provided $-8/5 \leqslant \lambda \leqslant 1/15$. Show also that the solution in this case is $x_2 = 4 - 3\lambda$, $x_3 = 5 - 6\lambda$, $x_6 = 11 - 9\lambda$ and that $z = -11 + 15\lambda$.

1.5.2 [3] Show that if A remains fixed and N changes to $N + \lambda M$ then the region within which λ can lie is determined by the inequalities $\gamma_i - \lambda\delta_i \geqslant 0$, $i = m + 1, \ldots, n$ where

$$c_x A^{-1} M = [\delta_{m+1}, \ldots, \delta_n].$$

1.5.3 [3] Show that if A and N change to $A + \lambda B$ and $N + \lambda M$ respectively then there will be no second degree term in (1.5.17) provided M is taken as a solution of $BA^{-1}M = 0$. Show further that in this case the θ_i in (1.5.18) are unaltered and the μ_i in (1.5.21) are elements of $c_x P$ where $P = A^{-1}BA^{-1}N - A^{-1}M$.

REFERENCES

1. Barnett, S., 'Stability of the solution to a linear programming problem', *Operat. Res. Quart.* **13**, 219–228 (1962).
2. Barnett, S., 'Polynomial matrices and a problem in linear system theory', *IEEE Trans. Aut. Control*, **AC-13**, 216–217 (1968).
3. Barnett, S., 'A simple class of parametric linear programming problems', *Opns. Res.* **16**, 1160–1165 (1968).
4. Barnett, S., 'Regular polynomial matrices having relatively prime determinants', *Proc. Cambridge Philos. Soc.* **65**, 585–590 (1969).

5. BARNETT, S., 'Degrees of greatest common divisors of invariant factors of two regular polynomial matrices', *Proc. Cambridge Philos. Soc.* **66,** 241–245 (1969).
6. BELEVITCH, V., 'On network analysis by polynomial matrices', in *Recent developments in network theory,* Ed. S. R. Deards, Pergamon Press, Oxford (1963).
7. BÔCHER, M., *Introduction to higher algebra,* Dover, New York (1964) (reprint of 1907 Edn.).
8. CHÂTELET, A., *Les groupes abéliens finis et les modules de points entiers,* Paris (1924).
9. COURTILLOT, M., 'On varying all the parameters in a linear-programming problem and sequential solutions of a linear programming problem', *Opns. Res.* **10,** 471–475 (1962).
10. CULLEN C. G., *Matrices and linear transformations,* Addison-Wesley, Reading, Mass. (1966).
11. DANTZIG, G. B., *Linear programming and extensions,* Princeton University Press, New Jersey (1963).
12. DEWEY, A. G., 'Computation of the Smith canonical form of a polynomial matrix', *Electronics Letters,* **3,** 122 (1967).
13. FLOOD, M. M., 'Division by non-singular matrix polynomials', *Ann. Math.* **36,** 859–869 (1935).
14. GANTMACHER, F. R., *Theory of matrices,* Vol. 1, Chelsea Publishing Company, New York (1960).
15. GASS, S. I., *Linear programming,* 2nd Edn., McGraw-Hill, New York (1964).
16. HODGE, W. V. D. and PEDOE, D., *Methods of algebraic geometry,* Vol. I, Cambridge University Press (1947).
17. JONES, J., JR., 'On the Lyapunov stability criteria', *SIAM J. Appl. Math.* **13,** 941–945 (1965).
18. KALMAN, R. E., 'Irreducible realizations and the degree of a rational matrix', *SIAM J. Appl. Math.* **13,** 520–544 (1965).
19. KALMAN, R. E., 'On structural properties of linear, constant, multivariable systems', *IFAC Conference,* London (1966).
20. KALMAN, R. E., FALB, P. L., and ARBIB, M. A., *Topics in mathematical system theory,* McGraw-Hill, New York (1969).
21. LANCASTER, P., *Lambda-matrices and vibrating systems,* Pergamon Press, Oxford (1966).
22. MACDUFFEE, C. C., *The theory of matrices,* Chelsea Publishing Company, New York (1956).
23. MACFARLANE, A. G. J., 'Multivariable-control-system design techniques: A guided tour', *Proc. IEE,* **117,** 1039–1047 (1970).
24. ROSENBROCK, H. H., 'On the design of linear multivariable control systems', *IFAC Conference,* London (1966).

25. ROSENBROCK, H. H., 'Transformation of linear constant system equations', *Proc. IEE*, **114,** 541–544 (1967).
26. ROSENBROCK, H. H., 'On linear system theory', *Proc. IEE*, **114,** 1353–1359 (1967).
27. ROSENBROCK, H. H., 'Least order of system matrices', *Electronics Letters*, **3,** 58–59 (1967).
28. ROSENBROCK, H. H., 'A connection between network theory and the theory of linear dynamical systems', *Electronics Letters*, **3,** 296–297 (1967).
29. ROSENBROCK, H. H., 'Reduction of system matrices', *Electronics Letters*, **3,** 368 (1967).
30. ROSENBROCK, H. H., 'Efficient computation of least order', *Electronics Letters*, **3,** 413–414 (1967).
31. ROSENBROCK, H. H., 'Generation of polynomial system matrices', *Electronics Letters*, **3,** 486–487 (1967).
32. ROSENBROCK, H. H., 'Relatively prime polynomial matrices', *Electronics Letters*, **4,** 227–228 (1968).
33. ROSENBROCK, H. H., 'Generalised resultant', *Electronics Letters*, **4,** 250–251 (1968).
34. ROSENBROCK, H. H., 'Some properties of relatively prime polynomial matrices', *Electronics Letters*, **4,** 374–375 (1968).
35. ROSENBROCK, H. H., 'Contributions to discussion on multivariable control systems', University of Manchester Institute of Science and Technology, Control Systems Centre Report No. 41, October, 1968.
36. ROSENBROCK, H. H., *State-space and multivariable theory*, Nelson, London (1970).
37. ROSENBROCK, H. H. and ROWE, A., 'Allocation of poles and zeros', *Proc. IEE*, **117,** 1879–1886 (1970).
38. ROTH, W. E., 'The equations $AX - YB = C$ and $AX - XB = C$ in matrices', *Proc. Amer. Math. Soc.* **3,** 392–396 (1952).
39. SAATY, T. L. and GASS, S. I., 'The parametric objective function, Part 1', *Opns. Res.* **2,** 316–319 (1954).
40. SHETTY, C. M., 'On analyses of the solution to a linear programming problem', *Operat. Res. Quart.* **12,** 89–104 (1961).
41. SMITH, H. J. S., 'On systems of linear indeterminate equations and congruences', *Philos. Trans. Roy. Soc. London*, **151,** 293–326 (1861).
42. WILLIAMS, A. C., 'Marginal values in linear programming', *SIAM J. Appl. Math.* **11,** 82–94 (1963).
43. WILLNER, L. B., 'On parametric linear programming', *SIAM J. Appl. Math.* **15,** 1253–1257 (1967).
44. ZADEH, L. A. and DESOER, C. A., *Linear system theory*, McGraw-Hill, New York (1963).

2

Polynomials

2.1 Resultant and discriminant

Problems connected with polynomials and polynomial equations have interested mathematicians for centuries, and there is in consequence a large body of literature on the subject. Research is still actively being carried on into numerical, algebraic and analytical aspects—see, for example, [26] for a comprehensive account with a full bibliography in the last-named area. Although much current work is concerned with devising new or improved computer algorithms for calculation of roots of polynomial equations†, nevertheless the earlier results on algebraic properties of polynomials are still of great importance. This has been emphasized by a recent revival of interest [16, 17, 21] in material which is found mainly in older textbooks. In this section we give a review of some of the classical theorems, which are largely based on determinants, in order to lessen this gap in the literature, and then show in the next section how equivalents of these results can be obtained in terms of matrices.

We first consider two polynomials

$$a(\lambda) = a_0\lambda^n + a_1\lambda^{n-1} + \cdots + a_n \tag{2.1.1}$$

and

$$b(\lambda) = b_0\lambda^m + b_1\lambda^{m-1} + \cdots + b_m, \tag{2.1.2}$$

where the a_i and b_j belong to some field of numbers, which we can take in this chapter as the field of complex numbers.

The *resultant* $R(a, b)$ of $a(\lambda)$ and $b(\lambda)$ is defined as follows:

---

† *The roots of a polynomial equation are referred to as the* zeros *of the polynomial itself.*

THEOREM 2.1 The polynomials $a(\lambda)$ and $b(\lambda)$ have a common factor (of degree greater than zero) if and only if the $(m+n)$ order *Sylvester determinant*

$$
R(a,b) = \begin{vmatrix}
a_0 & a_1 & a_2 & \cdots & \cdot & a_n & 0 & \cdot & \cdot \\
0 & a_0 & a_1 & \cdots & \cdot & a_{n-1} & a_n & \cdot & \cdot \\
\cdot & \cdot & \cdot & \cdots & \cdot & \cdot & \cdot & \cdot & \cdot \\
\cdot & \cdot & \cdot & \cdots & \cdot & \cdot & \cdot & a_{n-1} & a_n \\
\cdot & \cdot & \cdot & \cdots & b_0 & b_1 & \cdot & b_{m-1} & b_m \\
\cdot & \cdot & \cdot & \cdots b_0 & b_1 & b_2 & \cdot & b_m & 0 \\
\cdot & \cdot & \cdot & \cdots & \cdot & \cdot & \cdot & \cdot & \cdot \\
b_0 & b_1 & b_2 & \cdots & \cdot & \cdot & \cdot & \cdot & \cdot
\end{vmatrix}
\begin{matrix} \uparrow \\ m \\ \text{rows} \\ \downarrow \\ \uparrow \\ n \\ \text{rows} \\ \downarrow \end{matrix}
\qquad (2.1.3)
$$

is zero, provided a_0 and b_0 are not both zero.

We note the following:

(i) For a proof, see for example [10]. The treatment used there depends upon the result quoted at the beginning of Section 1.2. However, the necessity of the condition of Theorem 2.1 is easily demonstrated by the following argument, due to Sylvester. Consider the equations obtained by setting $a(\lambda)$ and $b(\lambda)$ in (2.1.1) and (2.1.2) equal to zero. If these equations have a common root (which is, of course, equivalent to the polynomials $a(\lambda)$ and $b(\lambda)$ having a common factor of degree greater than zero), then multiplying (2.1.1) successively by $\lambda^{m-1}, \ldots, \lambda, 1$ and (2.1.2) by $1, \lambda, \ldots, \lambda^{n-1}$ we obtain the set of $m+n$ equations:

$$
\begin{aligned}
a_0\lambda^{n+m-1} + a_1\lambda^{n+m-2} + \cdots \quad \cdot \quad \cdots + a_n\lambda^{m-1} &= 0 \\
a_0\lambda^{n+m-2} + \cdots \quad \cdot \quad \cdots + a_{n-1}\lambda^{m-1} + a_n\lambda^{m-2} &= 0 \\
\cdot \qquad \cdot \qquad \cdot \qquad \cdot \qquad \cdot & \\
\cdot \qquad \cdot \qquad \cdot \qquad \cdot \qquad \cdot & \\
a_0\lambda^n + a_1\lambda^{n-1} + \cdots \qquad \cdots + a_n &= 0 \\
b_0\lambda^m + b_1\lambda^{m-1} + \cdots + b_m &= 0 \\
\cdot \qquad \cdot \qquad \cdot \qquad \cdot \qquad \cdot & \\
\cdot \qquad \cdot \qquad \cdot \qquad \cdot \qquad \cdot & \\
b_0\lambda^{n+m-1} + b_1\lambda^{n+m-2} + \cdots + b_m\lambda^{n-1} &= 0
\end{aligned}
$$

Regarding these as linear homogeneous equations in λ, $\lambda^2, \ldots, \lambda^{m+n-1}$ the condition for consistency is just that $R(a, b)$ be zero. This procedure is known as the 'dialytic method of elimination', and the determinant thus obtained is termed the *eliminant* of the two equations.

(ii) Since we have taken polynomials over the field of complex numbers, it is possible to factorize $a(\lambda)$ and $b(\lambda)$ uniquely into linear factors:

$$a(\lambda) = a_0(\lambda - \alpha_1)(\lambda - \alpha_2)\cdots(\lambda - \alpha_n) \tag{2.1.4}$$

$$b(\lambda) = b_0(\lambda - \beta_1)(\lambda - \beta_2)\cdots(\lambda - \beta_m). \tag{2.1.5}$$

It is then easily shown [1] that

$$\begin{aligned} R(a, b) &= \varepsilon(n)a_0^m b_0^n \prod_{i=1}^{n}\prod_{j=1}^{m}(\alpha_i - \beta_j) \qquad (2.1.6)\\ &= \varepsilon(n)a_0^m b(\alpha_1)b(\alpha_2)\cdots b(\alpha_n)\\ &= \varepsilon(n)(-1)^{mn}b_0^n a(\beta_1)a(\beta_2)\cdots a(\beta_m), \end{aligned}$$

where† $\varepsilon(n) = (-1)^{n(n-1)/2}$.

(iii) An equivalent way of stating the theorem is that $a(\lambda)$ and $b(\lambda)$ are relatively prime if and only if the resultant is nonzero.

Next consider the single polynomial $a(\lambda)$ in (2.1.4). The α_i may or may not be all distinct, and a simple criterion is obtained from Theorem 2.1 by realizing that $a(\lambda)$ has a repeated factor (i.e. a factor of multiplicity greater than or equal to two) if and only if this factor also divides the derivative $a'(\lambda)$. With a slight modification we have:

THEOREM 2.2 The polynomial $a(\lambda)$ has a repeated factor if and only if the *discriminant*‡

$$D(a) = (1/a_0)R(a, a') \tag{2.1.7}$$

is zero.

It is easy to establish (see [1]) that

$$D(a) = a_0^{2n-2}\prod_{i<j}(\alpha_i - \alpha_j)^2. \tag{2.1.8}$$

Another formula for D which is also sometimes useful is the nth order determinant a_0^{2n-2} det S_n, where

$$S_k = \begin{bmatrix} s_0 & s_1 & s_2 & \cdots & s_{k-1}\\ s_1 & s_2 & s_3 & \cdots & s_k\\ s_2 & s_3 & s_4 & \cdots & s_{k+1}\\ \cdot & \cdot & \cdot & \cdots & \cdot\\ s_{k-1} & s_k & s_{k+1} & \cdots & s_{2k-2} \end{bmatrix}, \tag{2.1.9}$$

and where $s_i = \alpha_1^{\,i} + \alpha_2^{\,i} + \cdots + \alpha_n^{\,i}$. If A is a companion matrix of $a(\lambda)$ (see Appendix 1) then it is easy to calculate the s_i from $s_i = \text{tr}\,(A^i)$. The above determinant is also interesting because of the following two results.

THEOREM 2.3 The polynomial $a(\lambda)$ has exactly t distinct zeros if and only if

$$\det S_t \neq 0, \quad \det S_{t+1} = \det S_{t+2} = \cdots = \det S_n = 0.$$

† *Many authors define resultant so that the factor* $\varepsilon(n)$ *disappears; see Exercise* 2.1.5.

‡ *See Exercise* 2.1.5 *for an alternative definition.*

A proof of this theorem is given by Archbold [1, p. 407].

If we are interested in finding the number of distinct *real* zeros of $a(\lambda)$, then there is no loss of generality in restricting ourselves to the case when all the a_i in (2.1.1) are real. For suppose that $a(\lambda) = e(\lambda) + if(\lambda)$, where $e(\lambda)$ and $f(\lambda)$ have real coefficients. Every real zero of $a(\lambda)$ is also a real zero of both $e(\lambda)$ and $f(\lambda)$, so the real zeros of $a(\lambda)$ are the same as those of the g.c.d. of $e(\lambda)$ and $f(\lambda)$, which of course also has real coefficients. Then

THEOREM 2.4 The number of distinct real zeros of $a(\lambda)$, with all a_i real, is equal to $t - 2V$, where V is the number of variations in sign in the sequence

$$1, \det S_2, \det S_3, \ldots, \det S_t.$$

A proof, based on Hankel forms, can be found in [15, vol. II, p. 203].

For large values of m and n the resultant and discriminant as defined above become difficult to calculate, and one way of reducing the order of the determinants involved is due to Bézout. For (2.1.3) this results in a determinant of order n or m, whichever is the greater, and details may be found in [1]. Another reduction is outlined in [1] whereby, if $n \geqslant m$ and $c(\lambda)$ is the remainder when $a(\lambda)$ is divided by $b(\lambda)$, $R(a, b)$ is shown to be a multiple of $R(a, c)$.

The concept of resultants was usefully extended by Trudi in 1862†: define the *bigradient* (or *subresultant*) as

$$B\{(a)_i, (b)_j\} = \begin{vmatrix} a_0 & a_1 & a_2 & \cdots & a_{i+j-1} \\ 0 & a_0 & a_1 & \cdots & a_{i+j-2} \\ 0 & 0 & a_0 & \cdots & a_{i+j-3} \\ \cdot & \cdot & \cdot & \cdots & \cdot \\ 0 & b_0 & b_1 & \cdots & b_{i+j-2} \\ b_0 & b_1 & b_2 & \cdots & b_{i+j-1} \end{vmatrix},$$

where there are i rows containing the a's, j rows containing the b's, and $a_i = 0, i > n, b_j = 0, j > m$. The name 'bigradient' arose because of the pattern of the two sets of diagonal nonzero entries. Clearly

$$B\{(a)_m, (b)_n\} \equiv R(a, b).$$

We illustrate the formation of subresultants by an example with $n = 4$, $m = 2$:

† *For historical material on bigradients, resultants and indeed any aspect of the theory of determinants, the monumental work by Muir* [28] *is invaluable.*

$$\begin{vmatrix} a_0 & a_1 & a_2 & a_3 & a_4 & 0 \\ 0 & a_0 & a_1 & a_2 & a_3 & a_4 \\ 0 & 0 & 0 & b_0 & b_1 & b_2 \\ 0 & 0 & b_0 & b_1 & b_2 & 0 \\ 0 & b_0 & b_1 & b_2 & 0 & 0 \\ b_0 & b_1 & b_2 & 0 & 0 & 0 \end{vmatrix}. \tag{2.1.10}$$

The determinant of the whole array is $R(a, b)$; the determinants of the arrays within dashed lines are $B\{(a)_1, (b)_3\}$ and $B\{(a)_0, (b)_2\}$.

It is also possible to form bigradients which are themselves polynomials:

$$B\{(a)_i, (b)_j, \lambda\} = \begin{vmatrix} a_0 & a_1 & a_2 & \cdots & a_{i+j-2} & \lambda^{i-1}a(\lambda) \\ 0 & a_0 & a_1 & \cdots & a_{i+j-3} & \lambda^{i-2}a(\lambda) \\ \cdot & \cdot & \cdot & \cdots & \cdot & \cdot \\ 0 & b_0 & b_1 & \cdots & b_{i+j-3} & \lambda^{j-2}b(\lambda) \\ b_0 & b_1 & b_2 & \cdots & b_{i+j-2} & \lambda^{j-1}b(\lambda) \end{vmatrix}. \tag{2.1.11}$$

We can now state Trudi's result.

THEOREM 2.5 If

$$\left.\begin{aligned} B\{(a)_m, (b)_n\} &= B\{(a)_{m-1}, (b)_{n-1}\} = \cdots \\ \cdots &= B\{(a)_{m-j+1}, (b)_{n-j+1}\} = 0 \\ B\{(a)_{m-j}, (b)_{n-j}\} &\neq 0, \end{aligned}\right\} \tag{2.1.12}$$

then the greatest common divisor of $a(\lambda)$ and $b(\lambda)$ has degree j and is $B\{(a)_{m-j}, (b)_{n-j}, \lambda\}$. Conversely, if the g.c.d. of $a(\lambda)$ and $b(\lambda)$ has degree j then it is $B\{(a)_{m-j}, (b)_{n-j}, \lambda\}$, and (2.1.12) holds.

Householder [16] gives a simple proof of the theorem in an expository paper on bigradients and related topics.

As an example, suppose that in (2.1.10) $B\{(a)_2, (b)_4\}$ is zero but $B\{(a)_1, (b)_3\} \neq 0$. Then the g.c.d. of $a(\lambda)$ and $b(\lambda)$ has degree one and is given by

$$\begin{vmatrix} a_0 & a_1 & a_2 & a(\lambda) \\ 0 & 0 & b_0 & b(\lambda) \\ 0 & b_0 & b_1 & \lambda b(\lambda) \\ b_0 & b_1 & b_2 & \lambda^2 b(\lambda) \end{vmatrix}.$$

Thus the last column of the first non-vanishing bigradient is replaced by the appropriate column in (2.1.11). It is worth pointing out, however, that (2.1.11) does not necessarily give the greatest common divisor as a *monic* polynomial. For an alternative method of obtaining the g.c.d. from (2.1.3), see [22].

Resultants can be used to solve two simultaneous algebraic equations in two unknowns (see Exercise 2.1.4) and have been incorporated by Bareiss [3] in an algorithm for finding the zeros of a polynomial. Epstein [12] has applied resultants to the problem of constructing a polynomial $h(\lambda)$ whose zeros are the values assumed by a given polynomial $f(\lambda)$ at the zeros of a second given polynomial $g(\lambda)$. In particular he specializes to the case when $f(\lambda)$ has real coefficients and $g(\lambda) = f'(\lambda)$, thus obtaining the relative extreme values of $f(\lambda)$ as the zeros of $h(\lambda)$ (see also Exercise 2.2.2).

When more than two polynomials are considered the situation becomes more complicated. If there are three polynomials of the same degree than an eliminant which is a determinant can be found, but when the degrees are different the eliminant is the ratio of two determinants (recall that the eliminant gives only a necessary condition that the polynomials have a non-constant g.c.d.). This work of Sylvester and Cayley can only be found, so far as the author is aware, in older textbooks (e.g. [34] or [29]). A similar procedure is possible for more than three polynomials, but the manipulation needed is rather clumsy.

Further relationships involving bigradients, Hankel determinants, the Padé table and other topics are admirably set out in the two papers by Householder previously referred to and need not be repeated here. Other material can also be found in textbooks on the theory of equations (e.g. [35]) and in [27] and [13].

EXERCISES

2.1.1 Show that the discriminants of $a\lambda^2 + b\lambda + c$ and $\lambda^3 + 3b\lambda + c$ are $b^2 - 4ac$ and $-27(c^2 + 4b^3)$ respectively.

2.1.2 Show that

$$B\{(a)_i, (b)_j, \lambda\} = \alpha_i(\lambda)a(\lambda) + \beta_j(\lambda)b(\lambda),$$

where $\delta\{\alpha_i(\lambda)\} \leqslant i - 1$ and $\delta\{\beta_j(\lambda)\} \leqslant j - 1$.

2.1.3 Let

$$V = \begin{bmatrix} 1 & \alpha_1 & \dots & \alpha_1^{n-1} \\ 1 & \alpha_2 & \dots & \alpha_2^{n-1} \\ \cdot & \cdot & \dots & \cdot \\ 1 & \alpha_n & \dots & \alpha_n^{n-1} \end{bmatrix},$$

so det V is the *Vandermonde* determinant. Show from (2.1.8) that $D(a) = a_0^{2n-2} \det (V^T V)$ and hence derive (2.1.9).

2.1.4 Let $f(x, y) = x^2y + x^2 + 2xy + y^3$ and $g(x, y) = x^2 - 6x - 3y^2$. Regard f and g as polynomials in x with coefficients dependent on y and show that $R(f, g) = -y^4(4y + 3)^2$. Hence show that the solutions of the simultaneous equations $f = 0$, $g = 0$ are $(0, 0)$, $(3 + \frac{3}{4}\sqrt{19}, -\frac{3}{4})$ and $(3 - \frac{3}{4}\sqrt{19}, -\frac{3}{4})$.

2.1.5 Let $R(a, b) = \varepsilon(n)\rho(a, b)$ in (2.1.6). Hence show that if $a(\lambda)$, $b(\lambda)$ and $c(\lambda)$ are three polynomials, then $\rho(ab, c) = \rho(a, c)\rho(b, c)$.

Many authors (e.g. [1], [27]) use $\rho(a, b)$ as the definition of resultant, in which case $D(a) = (1/a_0)\varepsilon(n)\rho(a, a')$ and

$$\rho(a, b) = a_0^m b_0^n \prod_{i=1}^{n} \prod_{j=1}^{m} (\alpha_i - \beta_j).$$

2.1.6 Show that

$$D(ab) = D(a)D(b)\{R(a, b)\}^2.$$

2.1.7 Show that if $a(\lambda)$ has real coefficients and has no repeated zeros, then if $D(a) > 0$ there is an even number of pairs of complex conjugate zeros, and an odd number if $D(a) < 0$.

2.2 Matrix expressions

In this section we obtain matrix expressions for the resultant of two or more polynomials, their greatest common divisor and its degree, and for the discriminant of a polynomial. We begin by giving a result corresponding to Theorem 2.1; in this we required that a_0 and b_0 were not both zero, so for the whole of this section we shall assume that a_0 in (2.1.1) is unity. Let the companion matrix of $a(\lambda)$ be

$$A = \begin{bmatrix} 0 & 1 & 0 & \dots & 0 \\ 0 & 0 & 1 & \dots & 0 \\ \cdot & \cdot & \cdot & \dots & \cdot \\ 0 & 0 & 0 & \dots & 1 \\ -a_n & -a_{n-1} & -a_{n-2} & \dots & -a_1 \end{bmatrix} \tag{2.2.1}$$

(this is trivially different from (1.1.3)), so that $\det(\lambda I_n - A) = a(\lambda)$. Then

THEOREM 2.6 The polynomials $a(\lambda)$ and $b(\lambda)$ are relatively prime if and only if the $n \times n$ matrix

$$R = b(A) = b_0 A^m + b_1 A^{m-1} + \cdots + b_m I_n \tag{2.2.2}$$

is nonsingular.

Proof. The result follows at once by specializing Theorem 1.12. Alternatively, it is easy to obtain directly as follows. Since the characteristic roots of A are $\alpha_1, \alpha_2, \ldots, \alpha_n$, those of $b(A)$ are $b(\alpha_1), b(\alpha_2), \ldots, b(\alpha_n)$. Hence R is singular if and only if $b(\alpha_i)$ is zero for at least one i, which is equivalent to $a(\lambda)$ and $b(\lambda)$ having a common factor.

Notice also from (2.1.4) and (2.1.5) that

$$\begin{aligned} \det R &= b_0^n \det(A - \beta_1 I_n) \det(A - \beta_2 I_n) \cdots \det(A - \beta_m I_n) \\ &= b_0^n \prod_{i=1}^{n} \prod_{j=1}^{m} (\alpha_i - \beta_j) \end{aligned}$$

which, apart from a trivial factor, is the same as (2.1.6). Thus we have a matrix expression for resultant which, by interchanging the roles of

$a(\lambda)$ and $b(\lambda)$ if necessary, gives a determinant of order n or m whichever is the smaller, and so provides an improvement over the classical results of the previous section. MacDuffee [25] remarks that the expression in (2.2.2) is well known, but as far as the author is aware it has not appeared in any textbook, and in fact has been recently rediscovered independently by Kalman [20] and the author [4].

It is useful to develop at this stage a relation between the rows r_i $(i = 1, 2, \ldots, n)$ of R. For convenience take $m < n$; it is left as an easy exercise for the reader then to show that

$$r_1 = [b_m, b_{m-1}, \ldots, b_1, b_0, 0, \ldots, 0]. \tag{2.2.3}$$

If e_i is the ith row of I_n, then clearly $r_i = e_iR$. However, $e_i = e_{i-1}A$ so

$$\begin{aligned} r_i &= e_{i-1}AR \\ &= e_{i-1}RA \\ &= r_{i-1}A, \qquad i = 2, 3, \ldots, n. \end{aligned} \tag{2.2.4}$$

In fact the recurrence relation (2.2.4) between successive rows of R holds for any value of m, although if $m \geqslant n$ the expression (2.2.3) will of course be different. However, the case when $m = n$ is easily dealt with (see Exercise 2.2.5), so by interchanging $a(\lambda)$ and $b(\lambda)$ if $m > n$, equation (2.2.4) provides a simple method of constructing R in all possible cases.

The matrix R also gives the degree of the g.c.d. of $a(\lambda)$ and $b(\lambda)$, as we have seen in the previous chapter by specializing Theorem 1.13 (see Exercise 1.3.1). Thus:

THEOREM 2.7 The degree of the g.c.d. of $a(\lambda)$ and $b(\lambda)$ is $n - \text{rank } R$.

Proof. We can proceed directly by letting $b(\lambda) = b_0\Pi(\lambda - \gamma_i)^{\delta_i}$, the δ_i being the multiplicities of the factors of $b(\lambda)$. If $\mathcal{J}$ is the Jordan form of A then

$$\text{rank } b(A) = \text{rank } b_0\Pi(\mathcal{J} - \gamma_i I_n)^{\delta_i}.$$

The theorem then follows by using the fact that A is nonderogatory (so that there is only one Jordan block in $\mathcal{J}$ associated with each distinct characteristic root of A) and the details are left as an exercise for the reader.

Although also rediscovered by the author [4], the above theorem seems to have been first obtained by MacDuffee [25], using a somewhat different method of proof. However, what does seem to be new is the realization by the author that R can also be used to determine the g.c.d. itself, thus completing the correspondence with Theorem 2.5. Let

$$d(\lambda) = \lambda^k + d_1\lambda^{k-1} + \cdots + d_k$$

be the g.c.d. of $a(\lambda)$ and $b(\lambda)$. Then we have

THEOREM 2.8 [4] Let c_i denote the ith column of R in (2.2.2). Then $c_{k+1}, c_{k+2}, \ldots, c_n$ are linearly independent, and if

$$c_i = \sum_{j=k+1}^{n} x_{ij} c_j, \qquad i = 1, 2, \ldots, k, \tag{2.2.5}$$

then

$$d_p = x_{k+1-p,k+1}, \qquad p = 1, 2, \ldots, k.$$

Proof. First, let $b(\lambda) = b_1(\lambda)d(\lambda)$ and consider

$$d(A) = A^k + d_1 A^{k-1} + \cdots + d_k I_n.$$

By Theorem 2.7, $d(A)$ has rank $n - k$, and it is easily verified that the first $n - k$ rows of $d(A)$ are

$$\begin{bmatrix} d_k & d_{k-1} & \cdots & d_2 & d_1 & 1 & 0 & 0 & \cdots & 0 & 0 \\ 0 & d_k & \cdots & d_3 & d_2 & d_1 & 1 & 0 & \cdots & 0 & 0 \\ \cdot & \cdot & \cdots & \cdot & \cdot & \cdot & \cdot & \cdot & \cdots & \cdot & \cdot \\ 0 & \cdot & \cdots & \cdot & \cdot & \cdot & \cdot & \cdot & \cdots & d_1 & 1 \end{bmatrix}. \tag{2.2.6}$$

Clearly the last $n - k$ columns of $d(A)$ are linearly independent. If

$$s_i = \sum_{j=k+1}^{n} y_{ij} s_j, \qquad i = 1, 2, \ldots, k,$$

where s_i stands for the ith column of $d(A)$, then inspection of the first row of (2.2.6) gives $y_{i,k+1} = d_{k-i+1}$, $i = 1, 2, \ldots, k$.

Next, since $a(\lambda)$ and $b_1(\lambda)$ are relatively prime it follows by Theorem 2.6 that $b_1(A)$ is nonsingular. Since

$$\begin{aligned} c_i &= b_1(A) s_i \qquad (i = 1, 2, \ldots, n) \\ &= \sum_{j=k+1}^{n} y_{ij} b_1(A) s_j \\ &= \sum_{j=k+1}^{n} y_{ij} c_j \end{aligned}$$

it follows that $c_{k+1}, \ldots, c_n$ are linearly independent, and from (2.2.5) that $x_{ij} = y_{ij}$, all i, j. Thus in particular $x_{i,k+1} = d_{k-i+1}$, $i = 1, 2, \ldots, k$, and the desired result is obtained by setting $i = k + 1 - p$.

Again the matrix version is an improvement over the classical result, in this case especially so, since to calculate the required x's in Theorem 2.8 (by solving the set of linear equations (2.2.5)) is easier than expanding the determinant in Theorem 2.5. For example, let

$$a(\lambda) = \lambda^3 - 8\lambda^2 + 21\lambda - 18$$

and

$$b(\lambda) = \lambda^3 - \lambda^2 - 21\lambda + 45.$$

Then the matrix $R = b(A)$ is $A^3 - A^2 - 21A + 45I_3$ which, using Exercise 2.2.5 and equation (2.2.4), is

$$\begin{bmatrix} 63 & -42 & 7 \\ 126 & -84 & 14 \\ 252 & -168 & 28 \end{bmatrix}.$$

This has rank one, so by Theorem 2.7 the degree of the g.c.d. of $a(\lambda)$ and $b(\lambda)$ is two. Also, $c_1 = 9c_3$, $c_2 = -6c_3$, so by Theorem 2.8, $d_1 = -6$, $d_2 = 9$ and $d(\lambda) = \lambda^2 - 6\lambda + 9$.

Kalman [21] has also given a theorem on the g.c.d. of two polynomials in the context of his abstract approach to systems theory mentioned at the beginning of Section 4, Chapter 1, and Fryer [14] has shown how a Routh array can be used to obtain the g.c.d. (see also [36]).

We can now give a matrix formulation of the discriminant of a polynomial, corresponding to Theorems 2.2 and 2.3.

THEOREM 2.9 The polynomial $a(\lambda)$ has a repeated factor if and only if the matrix $a'(A)$ is nonsingular. Further, the number of distinct factors of $a(\lambda)$ is equal to the rank t of $a'(A)$.

Proof. The first part follows immediately from Theorem 2.6.

To prove the second part we follow MacDuffee [25]. Suppose $a(\lambda)$ has h_i distinct factors with multiplicity i, so that $n = \sum_{i=1}^{l} ih_i$.
The g.c.d. of $a(\lambda)$ and $a'(\lambda)$ thus has h_2 linear factors, h_3 with multiplicity two, etc. so that its degree is $\sum_{i=2}^{l} (i-1)h_i$. Hence, by Theorem 2.7,

$$\begin{aligned} \text{rank } a'(A) &= n - \sum_{i=2}^{l} (i-1)h_i \\ &= \sum_{i=1}^{l} h_i \qquad \text{as required.} \end{aligned}$$

It is interesting that the same matrix $a'(A)$ can also be used to determine the number of distinct *real* zeros, thus providing an alternative to Theorem 2.4:

THEOREM 2.10 [6] The number of distinct real zeros of $a(\lambda)$, with all a_i real, is equal to $t - 2V$, where V is the number of variations in sign in the sequence

$$1, \varepsilon(2)M_2, \varepsilon(3)M_3, \ldots, \varepsilon(t)M_t, \tag{2.2.7}$$

M_k is the minor of the first k rows and last k columns of $a'(A)$, and

$$\varepsilon(k) = (-1)^{k(k-1)/2}. \tag{2.2.8}$$

Proof. This consists of showing that $\varepsilon(k)M_k = \det S_k$, $k = 2, 3, \ldots, t$, where S_k is given by (2.1.9), and the result then follows immediately from Theorem 2.4.

First, denoting the rows of $a'(A)$ by $\gamma_1, \gamma_2, \ldots, \gamma_n$, we have, by virtue of (2.1.1), (2.2.3) and (2.2.4), (remembering that $a_0 = 1$),

$$\gamma_1 = [a_{n-1}, 2a_{n-2}, \ldots, (n-1)a_1, n] \tag{2.2.9}$$

and

$$\gamma_i = \gamma_{i-1}A, \qquad (i = 2, 3, \ldots, n). \tag{2.2.10}$$

Next, let

$$\mathcal{J} = \begin{bmatrix} 0 & 0 & \ldots & 0 & 1 \\ 0 & 0 & \ldots & 1 & 0 \\ \cdot & \cdot & \ldots & \cdot & \cdot \\ \cdot & 1 & \ldots & \cdot & \cdot \\ 1 & 0 & \ldots & 0 & 0 \end{bmatrix} \tag{2.2.11}$$

and let the matrix

$$C = a'(A)\mathcal{J} \tag{2.2.12}$$

have rows $c_1, c_2, \ldots, c_n$. It follows from (2.2.9) that

$$c_1 = [n, (n-1)a_1, \ldots, 2a_{n-2}, a_{n-1}] \tag{2.2.13}$$

and from (2.2.10) that

$$c_i = c_{i-1}B, \qquad (i = 2, 3, \ldots, n) \tag{2.2.14}$$

where

$$B = \mathcal{J}^{-1}A\mathcal{J}$$

$$= \begin{bmatrix} -a_1 & -a_2 & \ldots & \cdot & -a_n \\ 1 & 0 & \ldots & \cdot & 0 \\ 0 & 1 & \ldots & \cdot & \cdot \\ \cdot & \cdot & \ldots & \cdot & \cdot \\ 0 & 0 & \ldots & 1 & 0 \end{bmatrix}. \tag{2.2.15}$$

Defining also

$$U = \begin{bmatrix} 1 & a_1 & a_2 & \ldots & a_{n-1} \\ 0 & 1 & a_1 & \ldots & a_{n-2} \\ 0 & 0 & 1 & \ldots & a_{n-3} \\ \cdot & \cdot & \cdot & \ldots & \cdot \\ 0 & 0 & 0 & \ldots & 1 \end{bmatrix} \tag{2.2.16}$$

we now prove that

$$C = S_nU, \tag{2.2.17}$$

where S_n is given by (2.1.9). Let the rows of S_nU be $\omega_1, \omega_2, \ldots, \omega_n$. Then

$$\omega_1 = [s_0, a_1s_0 + s_1, a_2s_0 + a_1s_1 + s_2, \ldots, a_{n-1}s_0 + a_{n-2}s_1 + \cdots + a_1s_{n-2} + s_{n-1}],$$

and using Newton's formula [35]:

$$s_i + a_1s_{i-1} + \cdots + a_{i-1}s_1 + ia_i = 0, \qquad (i = 1, 2, \ldots, n-1;\ a_i = 0,\ i > n), \tag{2.2.18}$$

it follows that the jth element of ω_1, namely

$$a_{j-1}s_0 + a_{j-2}s_1 + \cdots + a_1 s_{j-2} + s_{j-1}, \qquad (j = 1, 2, \ldots, n)$$

is equal to $(n - j + 1)a_{j-1}$, which is just the jth element of c_1 in (2.2.13) so that $\omega_1 \equiv c_1$. Also,

$$\omega_i = [s_{i-1}, a_1 s_{i-1} + s_i, \ldots, a_{n-1}s_{i-1} + a_{n-2}s_i + \cdots + s_{n+i-2}],$$

and because of (2.2.14) we now only need to show that $w_i = w_{i-1}B$, $i = 2, 3, \ldots, n$ in order to establish (2.2.17). This follows easily using (2.2.15), (2.2.16) and (2.2.18), and is left as an exercise for the reader.

Finally, because of the form of U in (2.2.16), application of the Binet–Cauchy theorem (see, for example, [15] vol. I, p. 9) to (2.2.17) shows that the leading principal minor C_k (of order k) of C is equal to det S_k. Similarly, from (2.2.12), C_k is equal to $\varepsilon(k)M_k$, where $\varepsilon(k)$ is the minor of the first k columns and last k rows of J. We leave the derivation of (2.2.8) as another easy exercise for the reader, and this completes the proof.

As an example, take

$$a(\lambda) = \lambda^4 - \lambda^3 - \lambda + 1, \tag{2.2.19}$$

so that

$$A = \begin{bmatrix} 0 & 1 & 0 & 0 \\ 0 & 0 & 1 & 0 \\ 0 & 0 & 0 & 1 \\ -1 & 1 & 0 & 1 \end{bmatrix}.$$

From (2.2.9) and (2.2.10) we obtain

$$a'(A) = \begin{bmatrix} -1 & 0 & -3 & 4 \\ -4 & 3 & 0 & 1 \\ -1 & -3 & 3 & 1 \\ -1 & 0 & -3 & 4 \end{bmatrix}.$$

The rank of $a'(A)$ is 3, and the sequence (2.2.7) is 1, 3, -54 which contains one variation in sign, so $a(\lambda)$ has three distinct zeros, one of which is real.

Again we remark that the two preceding theorems are simpler to apply than corresponding classical results (Theorems 2.2, 2.3 and 2.4, or other approaches based on Sturm sequences, etc. [27]), particularly in view of the ease with which the matrix $a'(A)$ can be obtained from (2.2.9) and (2.2.10).

As an example of another application of the foregoing procedures, we now take the problem of finding an equation whose roots are the n squared differences $(\alpha_i - \alpha_j)^2$, $i > j$, where $\alpha_1, \alpha_2, \ldots, \alpha_n$ are the roots of $a(\lambda) = 0$ [25]. First consider an equation $f(\mu) = 0$, whose roots are the n^2 differences $\alpha_i - \alpha_j$, n of these being zero and the remainder occurring in pairs with opposite signs. Thus $f(\mu) = \mu^n g(\mu^2)$, and the roots of $g(\nu)$ are $(\alpha_i - \alpha_j)^2$, $i > j$, as required. The construction of $f(\mu)$ becomes clear when it is realized that

$$a(\lambda) = 0,\ a(\lambda + \mu) = 0$$

have a common root for every μ which is a root of $f(\mu) = 0$, for then $\alpha_i = \alpha_j + \mu_k$. In other words, $f(\mu)$ is the resultant of $a(\lambda)$ and $a(\lambda + \mu)$, so by Theorem 2.6,

$$f(\mu) = \det\{a(A + \mu I_n)\}. \tag{2.2.20}$$

MacDuffee [25] also applies a similar method to the theory of symmetric functions.

It is very interesting that when more than two polynomials are involved, the results concerning their g.c.d. can be generalized without much difficulty. This is in sharp contrast to the classical approach which, as we have mentioned, is not at all easy to extend.

Suppose we have $m + 1$ polynomials, and let one of these having highest degree n be denoted by $a(\lambda)$ as in (2.1.1) (again with $a_0 = 1$ for convenience). If any other of the polynomials also has degree n, it may be replaced by its remainder after division by $a(\lambda)$, since this will not affect the g.c.d. Without loss of generality we can therefore denote the remaining polynomials by

$$b_i(\lambda) = b_{i1}\lambda^{n-1} + b_{i2}\lambda^{n-2} + \cdots + b_{in}, \qquad (i = 1, 2, \ldots, m)$$

(the b_{i1} need not, of course, be nonzero).

THEOREM 2.11 The degree of the g.c.d. of $a(\lambda), b_1(\lambda), \ldots, b_m(\lambda)$ is $n - \operatorname{rank} P$, where

$$P = \begin{bmatrix} b_1(A) \\ b_2(A) \\ \cdot \\ \cdot \\ \cdot \\ b_m(A) \end{bmatrix}.$$

Proof. This can be achieved along similar lines to those indicated for Theorem 2.7, the extension being straightforward but tedious.

Alternatively the result follows as a special case of Theorem 1.13. First notice that the proof of Theorem 1.13 still holds even if $B(\lambda)$ is rectangular. We therefore consider equation (1.3.1) with C replaced by A and

$$B(\lambda) = \begin{bmatrix} b_1(\lambda) \\ b_2(\lambda) \\ \cdot \\ \cdot \\ \cdot \\ b_m(\lambda) \end{bmatrix} = B_1\lambda^{n-1} + B_2\lambda^{n-2} + \cdots + B_n$$

where

$$B_j = \begin{bmatrix} b_{1j} \\ \cdot \\ \cdot \\ \cdot \\ b_{mj} \end{bmatrix} \qquad (j = 1, 2, \ldots, n).$$

The matrix in (1.3.1) becomes

$$\begin{aligned} B_1 \otimes A^{n-1} + B_2 \otimes A^{n-2} + \cdots + B_n \otimes I_n \\ = \begin{bmatrix} b_{11}A^{n-1} + b_{12}A^{n-2} + \cdots + b_{1n}I_n \\ \cdot \\ \cdot \\ \cdot \\ b_{m1}A^{n-1} + b_{m-2}A^{n-2} + \cdots + b_{mn}I_n \end{bmatrix} \\ = P. \end{aligned}$$

Clearly $B(\lambda)$ has the single invariant factor $d(\lambda)$, the g.c.d. of $b_1(\lambda), \ldots, b_m(\lambda)$, and by Theorem 1.13 the degree of the g.c.d. of $a(\lambda)$ and $d(\lambda)$ is $n -$ rank P.

It has been pointed out by Rosenbrock† that Theorem 2.11 can also be deduced from the result in linear system theory [33, Chapter 2, Theorem 8.1] that if $e(\lambda)$ is the nth determinantal divisor of the matrix $E(\lambda) = \begin{bmatrix} \lambda I_n - A \\ B \end{bmatrix}$ then $\delta\{e(\lambda)\} = n -$ rank F, where

$$F = \begin{bmatrix} B \\ BA \\ \cdot \\ \cdot \\ \cdot \\ BA^{n-1} \end{bmatrix}. \tag{2.2.21}$$

To demonstrate the relationship, let B in (2.2.21) have rows $h_1, \ldots, h_m$ where $h_i = [b_{in}, \ldots, b_{i1}]$. Then by equivalence transformations like those suggested in Exercise 1.1.8 it is easy to reduce $E(\lambda)$ to

$$\begin{bmatrix} 1 & & & & \\ & 1 & & 0 & \\ & & \cdot & & \\ & & & \cdot & \\ & & & & \cdot \\ & & & & 1 \quad 0 \\ & & & & a(\lambda) \\ & & & & b_1(\lambda) \\ & & 0 & & \cdot \\ & & & & \cdot \\ & & & & \cdot \\ & & & & b_m(\lambda) \end{bmatrix},$$

† *Private communication to the author.*

showing that $e(\lambda)$ is the g.c.d. of $a(\lambda), b_1(\lambda), \ldots, b_m(\lambda)$. Also, from (2.2.3) and (2.2.4),

$$b_i(A) = \begin{bmatrix} h_i \\ h_iA \\ \cdot \\ \cdot \\ \cdot \\ h_iA^{n-1} \end{bmatrix}$$

(see Exercise 2.2.3), and simple rearrangement of the rows of F establishes that rank F = rank P.

We recognize F^T as the matrix involved in the controllability criterion (Appendix 2) for the pair $[A^T, B^T]$.

It is easy to verify that the result and proof of Theorem 2.8 on the g.c.d. itself carry over directly to the case of more than two polynomials. The only differences are that k = rank P and that in the expression (2.2.5) the c_i are now the columns of P. Thus the problem of finding the g.c.d. of several polynomials reduces to solving a set of linear equations of the form (2.2.5), and again this is more straightforward than any of the classical methods and seems a useful extension [8].

EXERCISES

2.2.1 [25] Let $a_1(\lambda)$ and $a_2(\lambda)$ be two relatively prime polynomials with companion matrices A_1 and A_2 respectively. Then the companion matrix of the polynomial $a(\lambda) = a_1(\lambda)a_2(\lambda)$ is similar to

$$\begin{bmatrix} A_1 & 0 \\ 0 & A_2 \end{bmatrix}.$$

Using this result, show that

$D(a) = D(a_1)D(a_2)\{R(a_1, a_2)\}^2$ (compare with Exercise 2.1.6).

2.2.2 If A' stands for the companion matrix of $a'(\lambda)$, show that the largest real characteristic root of $a(A')$ will be the relative maximum of the real polynomial $a(\lambda)$ (see also [12]—in fact it may be possible to solve fully the problem discussed there using the matrix techniques of this section).

2.2.3 Show that the resultant of $\lambda^n + a_1\lambda^{n-1} + \cdots + a_n$ and $b_1\lambda^{n-1} + b_2\lambda^{n-2} + \cdots + b_n$ can be written in the form $\det [h^T, A^Th^T, \ldots, (A^T)^{n-1}h^T]$, where $h = [b_n, b_{n-1}, \ldots, b_1]$. This is similar to the form in which Kalman [20] rediscovered the matrix expression for resultant. Notice the close relationship with controllability (see Appendix 2).

2.2.4 [25] Show that the equation of which the roots are the squared differences of the roots of $\lambda^3 - 7\lambda + 6 = 0$ is

$$\nu^3 - 42\nu^2 + 441\nu - 400 = 0.$$

Hence deduce the magnitude of the discriminant of $\lambda^3 - 7\lambda + 6$.

2.2.5 If $m = n$, use the Cayley–Hamilton theorem to show that the first row of R in (2.2.2) is $r_1 = (r_{1n}, r_{1,n-1}, \ldots, r_{11})$, where $r_{1i} = b_i - b_0 a_i$.

2.2.6 Show that the polynomial

$$\lambda^4 + 2\lambda^3 + 2\lambda^2 + 2\lambda + 1$$

has three distinct roots, of which one is real.

2.2.7 [27] Let $d_1(\lambda)$ be the g.c.d. of a given polynomial $d_0(\lambda)$ and its derivative $d_0'(\lambda)$, let $d_2(\lambda)$ be the g.c.d. of $d_1(\lambda)$ and $d_1'(\lambda)$ and so on, until we obtain a polynomial $d_s(\lambda)$ of zero degree. Let

$$v_{i+1}(\lambda) = d_i(\lambda)/d_{i+1}(\lambda), \quad (i = 0, 1, 2, \ldots, s - 1)$$

and

$$F_i(\lambda) = v_i(\lambda)/v_{i+1}(\lambda), \; i = 1, 2, \ldots, s \; \{v_{s+1}(\lambda) \equiv 1\}.$$

Show that $F_1(\lambda), \ldots, F_s(\lambda)$ have no repeated zeros and that the zeros of $F_i(\lambda)$ are all the zeros of $d_0(\lambda)$ of multiplicity i.

This method could be applied using Theorem 2.8.

2.2.8 Theorem 2.10 establishes indirectly that det $a'(A) = \varepsilon(n)D(a)$, where $D(a)$ is given by (2.1.8). Prove this directly.

2.3 Location of zeros of a complex polynomial

A fundamental problem in linear control theory and many other areas of applied mathematics is that of finding conditions for the zeros of the polynomial

$$\lambda^n + a_1\lambda^{n-1} + a_2\lambda^{n-2} + \cdots + a_{n-1}\lambda + a_n \qquad (2.3.1)$$

all to have negative real parts, so that the associated nth order homogeneous linear differential equation with constant coefficients is asymptotically stable (see Theorem 4.1). Polynomials having this property are often termed *Hurwitz* polynomials. This problem is dealt with fully in many textbooks (for example, [15] and [23] contain excellent and comprehensive accounts) and there is no value in reproducing here details of the well-established techniques. We can usefully mention, however, two papers [2, 30] which summarize many of the known criteria, and two texts which contain valuable analytic material [26, 37]. The best-known theorems are probably those due to Routh (1877) and Hurwitz (1895), but in fact the problem was first solved by Hermite in 1854. As might be expected, there are many interconnections between the various results, and Anderson [2] gives an account of these in the case when all the a_i in (2.3.1) are real.

We will consider the stability problem again at the end of this section, but we shall first be concerned with a more general version of the problem, namely to find how many zeros of the polynomial $a(\lambda)$ in (2.3.1), with all a_i complex, lie in the half-planes to the left and right of the imaginary axis. If the problem can be solved for any particular half-plane then it is of course easy to solve it for any other, using a simple transformation. The problem as stated above was also solved by Hermite, and we now present his result in a form due to Lehnigk [23, 24]. Associate with (2.3.1) the real symmetric $n \times n$ *Hermite* matrix $H = [h_{\mu\nu}]$ with

$$h_{\mu\nu} = (\mu - 1, \nu) + (\mu - 2, \nu + 1) + \cdots \\ \cdots + (1, \mu + \nu - 2) + (0, \mu + \nu - 1), \qquad \mu \leqslant \nu \tag{2.3.2}$$

and

$$(k, l) = 0 \text{ if } l > n$$
$$(k, l) = i^{k+l+1}[(-1)^l a_k \bar{a}_l + (-1)^{k+1} \bar{a}_k a_l], \\ k = 0, 1, \ldots, n-1; k < l \leqslant n; a_0 = 1. \tag{2.3.3}$$

For example,

$$h_{11} = (0, 1) = a_1 + \bar{a}_1,$$
$$h_{12}\,(= h_{21}) = (0, 2) = i(a_2 - \bar{a}_2)$$
$$h_{22} = (1, 2) + (0, 3)$$
$$= (a_1\bar{a}_2 + \bar{a}_1 a_2) - (a_3 + \bar{a}_3).$$

It should be noticed that the formula (2.3.2) can still be used even if $\mu > \nu$ provided (k, l) in (2.3.3) is interpreted as $-(l, k)$ for $k > l$, and zero for $k = l$.

Denoting the principal minors det $[h_{\mu\nu}]$, $\mu, \nu = 1, 2, \ldots, k$, of H by H_k, we have

THEOREM 2.12 (Hermite) Let the numbers $H_1, H_2, \ldots, H_n$ be different from zero. Then $a(\lambda)$ has p zeros with positive real parts and $n - p$ zeros with negative real parts, where p is the number of variations in sign in the sequence $1, H_1, H_2, \ldots, H_n$.

A proof of Hermite's theorem may be found in the book by Lehnigk [23], together with details of the rather complicated situation when some of the H_i are zero. Incidentally, Lehnigk has also given a prcof [24] of Theorem 2.12 using Liapunov theory (see Chapter 4, Section 2).

In the paper [16] referred to in the first section of this chapter, Householder demonstrates how bigradients can be used to solve this same problem†. Now we have already seen that the matrix R in equation (2.2.2) can be used instead of bigradients to provide exactly the same information concerning the g.c.d. of two polynomials, so it is natural to ask whether the correspondence can be continued by applying R to the problem of location of zeros of a complex polynomial. We now show that this is indeed the case.

It turns out to be convenient to introduce a second complex polynomial

$$f(\theta) = \theta^n + f_1\theta^{n-1} + \cdots + f_n \tag{2.3.4}$$

† *See* [18] *for an exposition of the application of Bezoutiants.*

and to relate it to $a(\lambda)$ by putting $\theta = i\lambda$ in (2.3.4) and setting

$$f(i\lambda) = i^n a(\lambda). \tag{2.3.5}$$

Thus from (2.3.1), (2.3.4) and (2.3.5) we have

$$f_k = i^k a_k, \qquad k = 1, 2, \ldots, n. \tag{2.3.6}$$

Also, let $f_k = \phi_k + i\psi_k$, $k = 1, 2, \ldots, n$, where ϕ_k and ψ_k are real, so that

$$\begin{aligned} f(\theta) &= \theta^n + \phi_1\theta^{n-1} + \cdots + \phi_n + i(\psi_1\theta^{n-1} + \cdots + \psi_n) \\ &= \phi(\theta) + i\psi(\theta). \end{aligned} \tag{2.3.7}$$

Using (2.3.6) and (2.3.7), the relationship (2.3.3) for the Hermite matrix associated with $a(\lambda)$ becomes

$$\begin{aligned} (k, l) &= i(f_k\bar{f}_l - \bar{f}_k f_l) \\ &= 2(\phi_k\psi_l - \phi_l\psi_k). \end{aligned} \tag{2.3.8}$$

If we denote the rows of H by h_r $(r = 1, 2, \ldots, n)$, then equations (2.3.2) and (2.3.8) give

$$h_1 = 2[\psi_1, \psi_2, \ldots, \psi_n] \tag{2.3.9}$$

and

$$\begin{aligned} h_{rs} - h_{r-1,s+1} &= (r-1, s) \\ &= 2(\phi_{r-1}\psi_s - \phi_s\psi_{r-1}) \\ &= \phi_{r-1}h_{1s} - \phi_s h_{r-1,r1}, \end{aligned}$$

so there is a relation between the rows of H which can be written

$$h_r = \phi_{r-1}h_1 + h_{r-1}E, \qquad r = 2, 3, \ldots, n, \tag{2.3.10}$$

where

$$E = \begin{bmatrix} -\phi_1 & -\phi_2 & \cdots & \cdot & -\phi_n \\ 1 & 0 & \cdots & \cdot & 0 \\ 0 & 1 & \cdots & \cdot & 0 \\ \cdot & \cdot & \cdots & \cdot & \cdot \\ 0 & 0 & \cdots & 1 & 0 \end{bmatrix}.$$

Let F be the companion matrix in the form (2.2.1) for the real polynomial $\phi(\theta)$, let $R(f) = \psi(F)$ and let R_k $(k = 1, 2, \ldots, n)$ denote the $k \times k$ minor formed from the first k rows and last k columns of $R(f)$. We can now state an alternative to Theorem 2.12.

THEOREM 2.13 [5] Suppose none of $R_1, R_2, \ldots, R_n$ is zero. Then $a(\lambda)$ has p zeros with positive real parts and $n - p$ zeros with negative real parts, where p is the number of variations in sign in the sequence

$$1, R_1, \varepsilon(2)R_2, \ldots, \varepsilon(n)R_n.$$

Proof. This follows similar lines to the proof of Theorem 2.10, and consists of showing that $(\frac{1}{2})^k H_k = \varepsilon(k)R_k$, $k = 1, 2, \ldots, n$, where the H_k are as in Theorem 2.12, and $\varepsilon(k)$ is given by (2.2.8).

First, put

$$S = R(f)J, \tag{2.3.11}$$

where J is given by (2.2.11). If the rows of $R(f)$ are denoted by $\rho_1, \ldots, \rho_n$, then from (2.2.3) and (2.2.4) we have $\rho_1 = [\psi_n, \psi_{n-1}, \ldots, \psi_1]$ and $\rho_i = \rho_{i-1}F$. Hence, if the rows of S are $\tau_1, \ldots, \tau_n$, then from (2.3.11) we have $\tau_1 = [\psi_1, \psi_2, \ldots, \psi_n]$ and $\tau_i = \rho_i J = \rho_{i-1}FJ = \tau_{i-1}J^{-1}FJ$. It is then easy to verify that

$$\tau_i = \tau_{i-1}E. \tag{2.3.12}$$

Next, let

$$W = \begin{bmatrix} 1 & 0 & 0 & \cdots & \cdot & \cdot \\ \phi_1 & 1 & 0 & \cdots & \cdot & \cdot \\ \phi_2 & \phi_1 & 1 & \cdots & \cdot & \cdot \\ \cdot & \cdot & \cdot & \cdots & \cdot & \cdot \\ \cdot & \cdot & \cdot & \cdots & 1 & 0 \\ \phi_{n-1} & \phi_{n-2} & \phi_{n-3} & \cdots & \phi_1 & 1 \end{bmatrix},$$

and consider the product WS. If the rows of WS are $v_1, \ldots, v_n$, then

$$\begin{aligned} v_r &= \phi_{r-1}\tau_1 + \phi_{r-2}\tau_2 + \cdots + \tau_r \\ &= \tau_1(\phi_{r-1}I + \phi_{r-2}E + \phi_{r-3}E^2 + \cdots + E^{r-1}) \text{ by (2.3.12)} \\ &= \phi_{r-1}\tau_1 + v_{r-1}E, \qquad (r = 2, 3, \ldots, n). \end{aligned} \tag{2.3.13}$$

Since $v_1 = \tau_1 = \frac{1}{2}h_1$ (by (2.3.9)), comparison of (2.3.10) and (2.3.13) shows that

$$WS = \tfrac{1}{2}H. \tag{2.3.14}$$

The theorem then follows by applying the Binet–Cauchy result to (2.3.11) and (2.3.14), the details being very similar to those in the proof of Theorem 2.10.

Apart from its intrinsic mathematical interest, Theorem 2.13 is easier to apply than Hermite's result because of the simple way in which $R(f)$ can be obtained.

As an example, consider

$$a(\lambda) = \lambda^3 + \lambda^2 - 10\lambda + 8.$$

From (2.3.6) we have

$$\begin{aligned} f(\theta) &= \theta^3 + i\theta^2 + 10\theta - 8i \\ &= \theta^3 + 10\theta + i(\theta^2 - 8). \end{aligned}$$

Hence

$$F = \begin{bmatrix} 0 & 1 & 0 \\ 0 & 0 & 1 \\ 0 & -10 & 0 \end{bmatrix}$$

and $R(f) = F^2 - 8I_3$ is obtained from (2.2.3) and (2.2.4) as

$$\begin{bmatrix} -8 & 0 & 1 \\ 0 & -18 & 0 \\ 0 & 0 & -18 \end{bmatrix}.$$

The sequence in Theorem 2.13 is 1, 1, -18, 2592, so that $p = 2$. In fact $a(\lambda) = (\lambda - 1)(\lambda - 2)(\lambda + 4)$.

It is easy to show (see Exercise 2.3.2) that the condition $R_n \neq 0$ ensures that $a(\lambda)$ has no purely imaginary zero. However, it should be realized that even if $a(\lambda)$ does have no pure imaginary zero, R_n (and other R_i) may be zero (see Exercise 2.3.3). In this case since Theorem 2.13 is derived from Hermite's result, the theory which has been developed to deal with the situation when some of the H_i in Theorem 2.12 are zero [23] carries over directly.

Closely related to the problem of finding how many zeros of a polynomial lie in a given half-plane is that of finding how many lie inside the unit circle, since the transformation

$$\frac{\lambda + 1}{\lambda - 1} = \eta, \qquad \frac{\eta + 1}{\eta - 1} = \lambda \tag{2.3.15}$$

is a one-to-one mapping of the half-plane Re $(\lambda) < 0$ into the interior of the unit circle $|\eta| < 1$. This second problem also has important applications in its own right. For example, the solution of a linear difference equation with constant coefficients is asymptotically stable if and only if all the zeros of the associated polynomial lie inside the unit circle (see Theorem 4.7).

Let $a(\lambda)$ be as in (2.1.1) and define

$$g(\eta) = 2^{-n/2}(\eta - 1)^n a\left(\frac{\eta + 1}{\eta - 1}\right), \tag{2.3.16}$$

a polynomial of degree n. Clearly we have

$$g(\eta) = 2^{-n/2} \sum_{i=0}^{n} a_i(\eta + 1)^{n-i}(\eta - 1)^i \tag{2.3.17}$$

$$= \sum_{i=0}^{n} g_i \eta^{n-i}. \tag{2.3.18}$$

Following a recent paper by Duffin [11] we can give a simple expression for constructing a matrix $\Gamma = [\Gamma_{ij}]$, $i, j = 0, 1, \ldots, n$ which gives the relationship between the a_i and g_i. We have

THEOREM 2.14

$$[a_n, a_{n-1}, \ldots, a_1, a_0]\Gamma = [g_n, g_{n-1}, \ldots, g_1, g_0], \tag{2.3.19}$$

where

$$\Gamma_{ij} = \Gamma_{i,j+1} + \Gamma_{i-1,j} + \Gamma_{i-1,j+1}, \tag{2.3.20}$$

and if $\gamma_{ij} = 2^{n/2}\Gamma_{ij}$, the last column of the matrix $[\gamma_{ij}]$ has all unit entries, and the elements in the first row of $[\gamma_{ij}]$ are the binomial coefficients in the expansion of $(\eta - 1)^n$ in reverse order. Further, Γ is idempotent.

Proof. Equating terms containing a_{n-i} in (2.3.17) and (2.3.18) leads to

$$(\eta + 1)^i(\eta - 1)^{n-i} = \sum_{k=0}^{n} \Gamma_{ik}\eta^k.$$

Hence

$$(\eta + 1)\sum_{k=0}^{n} \Gamma_{ik}\eta^k = (\eta - 1)\sum_{k=0}^{n} \Gamma_{i+1,k}\eta^k. \qquad (2.3.21)$$

Equating coefficients of powers of η in (2.3.21) then readily produces (2.3.20).

It also follows directly from (2.3.17) and (2.3.18) that

$$g_0 = 2^{-n/2}\sum_{i=0}^{n} a_i,$$

so the elements of the last column of $[\gamma_{ij}]$ are all unity. The first row of Γ has the stated form since the term in a_n in (2.3.17) is $2^{-n/2}(\eta - 1)^n$.

From (2.3.15) and (2.3.16) we have

$$a(\lambda) = 2^{-n/2}(\lambda - 1)^n g\left(\frac{\lambda + 1}{\lambda - 1}\right),$$

so the preceding argument implies

$$[g_n, g_{n-1}, \ldots, g_0]\Gamma = [a_n, a_{n-1}, \ldots, a_1, a_0]. \qquad (2.3.22)$$

From (2.3.19) and (2.3.22) it then follows that $\Gamma^2 = I$, since the a_i are arbitrary.

The number of zeros of $a(\lambda)$ in the left half-plane is equal to the number of zeros of $g(\eta)$ inside the unit circle, so to test $g(\eta)$ the Hermite or other such algorithm could be applied to the corresponding $a(\lambda)$ found from (2.3.22). The matrix Γ is very easily obtained from (2.3.20); for example, when $n = 3$,

$$[\gamma_{ij}] = \begin{bmatrix} -1 & 3 & -3 & 1 \\ 1 & -1 & -1 & 1 \\ -1 & -1 & 1 & 1 \\ 1 & 3 & 3 & 1 \end{bmatrix}.$$

However, there do exist methods for determining directly the number of zeros of $g(\eta)$ inside $|\eta| < 1$, so that $a(\lambda)$ could be investigated by transforming through (2.3.19). The earliest solution to this direct problem is due to Schur and Cohn, and details can be found in [26]. We state the result here in a form similar to that quoted by Parks [30]:

THEOREM 2.15 Define the Hermitian matrix $K = [k_{ij}]$, $i, j = 0, 1, \ldots, n - 1$ by

$$k_{ij} = \sum_{r=0}^{i} (g_{i-r}\bar{g}_{j-r} - \bar{g}_{n+r-i}g_{n+r-j}), \qquad i \leqslant j \qquad (2.3.23)$$

and let K_i be its ith principal minor. If no K_i is zero then the polynomial

$$g(\eta) = \eta^n + g_1\eta^{n-1} + \cdots + g_n$$

has p zeros outside $|\eta| < 1$ and $n - p$ inside, where p is the number of variations in sign in the sequence $1, K_1, K_2, \ldots, K_n$.

A simplified procedure for computational purposes has been developed by Jury [19], and Theorem 2.15 has also been proved using Liapunov theory [31]. What is of interest to us here, however, is that the matrix methods we have developed can again be applied, and an alternative to Theorem 2.15 obtained.

Let

$$\tilde{g}(\eta) = \eta^n\left\{\bar{g}\left(\frac{1}{\eta}\right)\right\}$$

$$= \bar{g}_n\eta^n + \bar{g}_{n-1}\eta^{n-1} + \cdots + 1,$$

and let G be the companion matrix of $g(\eta)$ in the form (2.2.1). Then:

THEOREM 2.16 [7] The principal minors of $\tilde{g}(G)$ are equal to those of K.

Proof. This follows similar lines to the proofs of Theorems 2.10 and 2.13, and we therefore give only a brief outline here, and invite the reader to fill in the details.

The first row of $\tilde{g}(G)$ is

$$\gamma_1 = [1 - \bar{g}_n g_n, \bar{g}_1 - \bar{g}_n g_{n-1}, \ldots, \bar{g}_{n-1} - \bar{g}_n g_1]$$

(see Exercise 2.2.5) and the other rows are $\gamma_1 G, \gamma_1 G^2, \ldots, \gamma_1 G^{n-1}$. Defining

$$V = \begin{bmatrix} 1 & 0 & \cdots & 0 \\ g_1 & 1 & \cdots & 0 \\ g_2 & g_1 & \cdots & 0 \\ \cdot & \cdot & \cdots & \cdot \\ g_{n-1} & g_{n-2} & \cdots & 1 \end{bmatrix},$$

it is not difficult to show that $V\tilde{g}(G) = K$. This is done by establishing that the rows $v_1, v_2, \ldots, v_n$ of $V\tilde{g}(G)$ satisfy the recurrence relationship

$$v_i = g_{i-1}\gamma_1 + v_{i-1}G, \qquad (i = 2, 3, \ldots, n),$$

and that if $k_1, k_2, \ldots, k_n$ are the rows of K, then $k_1 = v_1$ and the k_i satisfy the same recurrence formula (this involves (2.3.23) and some rather tedious algebraic manipulation). The theorem then follows at once.

Once again it is easier to construct the matrix $\tilde{g}(G)$ than the Schur–Cohn matrix K. For the more complicated situation when some minors are zero (e.g. when there are zeros symmetric in the unit circle) see [26].

We close this chapter by returning to the case where the a_i in (2.3.1) are all real and all the zeros of $a(\lambda)$ are required to have negative real parts. One very well known necessary and sufficient condition is that the *Hurwitz determinants*

$$\Delta_i = \begin{vmatrix} a_1 & a_3 & a_5 & \cdots & a_{2i-1} \\ 1 & a_2 & a_4 & \cdots & a_{2i-2} \\ 0 & a_1 & a_3 & \cdots & a_{2i-3} \\ 0 & 1 & a_2 & \cdots & a_{2i-4} \\ \cdot & \cdot & \cdot & \cdots & \cdot \\ 0 & 0 & \cdot & \cdots & a_i \end{vmatrix}, \quad a_r = 0, r > n, \qquad (2.3.24)$$

be positive for $i = 1, 2, \ldots, n$. It is, however, easy to show that a necessary condition for stability is that all the a_i's be positive, and if this is indeed the case then only about half the Hurwitz determinants need to be calculated. This is because of a result due to Liénard and Chipart, quoted here in a form in which it was rediscovered by Fuller [13]:

THEOREM 2.17 Necessary and sufficient conditions for $a(\lambda)$ to be stable are that

$$a_n > 0, a_1 > 0, a_3 > 0, a_5 > 0, \ldots \qquad (2.3.25)$$

and

$$\Delta_{n-1} > 0, \Delta_{n-3} > 0, \Delta_{n-5} > 0, \ldots, \begin{Bmatrix} \Delta_3 > 0 \ (n \text{ even}) \\ \Delta_2 > 0 \ (n \text{ odd}) \end{Bmatrix}. \qquad (2.3.26)$$

Alternative sequences to (2.3.25) and (2.3.26) are possible and a full statement and proof of the Liénard–Chipart stability criteria can be found in [15, vol. II, p. 221]. Fuller's derivation of Theorem 2.17 is interesting because it relies on forming the resultant and subresultants of the two polynomials

$$c(\lambda) = a_n + a_{n-2}\lambda + a_{n-4}\lambda^2 + \cdots \qquad (2.3.27)$$

$$d(\lambda) = a_{n-1} + a_{n-3}\lambda + a_{n-5}\lambda^2 + \cdots \qquad (2.3.28)$$

In view of our previous results it is not surprising that the companion matrix approach once again provides a simplification:

THEOREM 2.18 [9] In Theorem 2.17, the sequence (2.3.26) is equivalent to the following:

(i) when n is even, the sequence

$$\varepsilon(k)D_k, \qquad k = 2, 3, \ldots, n/2 \qquad (2.3.29)$$

where D_k is the minor of the first k rows and last k columns of the matrix $d(C)$, and C is the companion matrix of $c(\lambda)$ in the form (2.2.1).

(ii) when n is odd, the sequence

$$\varepsilon(k+1)C_k, \qquad k = 1, 2, \ldots, (n-1)/2 \qquad (2.3.30)$$

where C_k is the minor of the first k rows and last k columns of $c(D)$ and D is the companion matrix of $d(\lambda)$ in the form (2.2.1).

In both cases $\varepsilon(k)$ is given by (2.2.8).

The key to the proof of the preceding theorem lies in a recently established relationship [8] between Sylvester's array (2.1.3) and the corresponding matrix polynomial $b(A)$. The details are straightforward but somewhat tedious and the reader is referred to [9].

When n is even $c(\lambda)$ in (2.3.27) is monic and has degree $n/2$, when n is odd $d(\lambda)$ in (2.3.28) is monic and has degree $(n-1)/2$. Thus the order of the largest determinant which has to be calculated in Theorem 2.18 is only about half that in Theorem 2.17. Furthermore, the matrices $d(C)$ and $c(D)$ are easy to calculate using (2.2.3), (2.2.4) and Exercise 2.2.5. It is worth pointing out also that the factors $\varepsilon(k)$ can be avoided if so desired, since (2.3.29) is the sequence of principal minors of $d(C)J_{n/2}$ (where J_n is defined in (2.2.11)), which is just the matrix formed by reversing the order of the columns of $d(C)$. Similarly (2.3.30) can be replaced by $(-1)^k$ times the principal minors of $c(D)J_{(n-1)/2}$.

As a simple example take

$$a(\lambda) = \lambda^7 + 4\lambda^6 + 10\lambda^5 + 16\lambda^4 + 18\lambda^3 + 14\lambda^2 + 7\lambda + 2, \tag{2.3.31}$$

so from (2.3.27) and (2.3.28)

$$c(\lambda) = 4\lambda^3 + 16\lambda^2 + 14\lambda + 2, \qquad d(\lambda) = \lambda^3 + 10\lambda^2 + 18\lambda + 7$$

and

$$D = \begin{bmatrix} 0 & 1 & 0 \\ 0 & 0 & 1 \\ -7 & -18 & -10 \end{bmatrix}.$$

Use of (2.2.4) and Exercise 2.2.5 produces

$$c(D) = \begin{bmatrix} -26 & -58 & -24 \\ 168 & 406 & 182 \\ -1274 & -3108 & -144 \end{bmatrix}$$

and it is easily verified that the sequence (2.3.30) is 24, 812, 7056. Since all the terms in (2.3.25) are also positive, (2.3.31) is a Hurwitz polynomial.

This simplified stability criterion completes our derivation of alternatives for many of the classical theorems in qualitative analysis of polynomials. Thus, *using matrices*, we can determine how many zeros of a given polynomial (a) are distinct (b) are real and distinct (c) lie in a given half-plane or inside the unit circle, without explicit calculation of the zeros. The methods used provide an interesting unification of the classical results, as well as some simplification of procedure. They may perhaps also provide a basis for development of such problems as sensitivity of the zeros of a polynomial to changes in its coefficients. In addition, it seems likely that further algebraic relationships involving classical results remain to be discovered.

EXERCISES

2.3.1 Use Theorems 2.12 and 2.13 to determine the location of the zeros of the polynomial $\lambda^3 + 4\lambda^2 + \lambda - 6$.

2.3.2 Show that relative primeness of $\phi(\theta)$ and $\psi(\theta)$ in (2.3.7) is a sufficient condition for $a(\lambda)$ in Theorem 2.13 to have no pure imaginary zero. Construct a simple example to show that the condition is not necessary.

2.3.3 Show that if $a(\lambda)$ has zeros z and $-\bar{z}$ where z is a complex number, then R_n in Theorem 2.13 is zero.

2.3.4 Parks [32] has given a simple proof of the known fact that for a *real* polynomial $a(\lambda)$, the minors H_i of the Hermite matrix are $2\Delta_1, 4\Delta_1\Delta_2, 8\Delta_2\Delta_3, \ldots, 2^n\Delta_{n-1}\Delta_n$, where Δ_i is given by (2.3.24). Verify this for the cases $n = 3, 4$ by finding the minors in Theorem 2.13.

2.3.5 Show that if $g(\eta)$ in Theorem 2.15 has zeros on or symmetric in the unit circle (i.e. $re^{i\theta}$ and $r^{-1}e^{i\theta}$), then K_n is zero.

REFERENCES

1. ARCHBOLD, A., *Algebra*, 3rd Edn., Pitman, London (1964).
2. ANDERSON, B. D. O., 'Application of the second method of Lyapunov to the proof of the Markov stability criterion', *Int. J. Control*, **5**, 473–482 (1967).
3. BAREISS, E. H., 'Resultant procedure and the mechanization of the Graeffe process', *Jnl. Assoc. for comp. mchry.* **7**, 346–386 (1960).
4. BARNETT, S., 'Greatest common divisor of two polynomials', *Linear Algebra and its Applications*, **3**, 7–9 (1970).
5. BARNETT, S., 'Location of zeros of a complex polynomial', *Linear Algebra and its Applications*, **4**, 71–76 (1971).
6. BARNETT, S., 'Qualitative analysis of polynomials using matrices', *IEEE Trans. Aut. Control*, **AC-15**, 380–382 (1970).
7. BARNETT, S., 'Number of zeros of a complex polynomial inside the unit circle', *Electronics Letters*, **6**, 164–165 (1970).
8. BARNETT, S., 'Greatest common divisor of several polynomials', *Proc. Cambridge Philos. Soc.* **70** (1971).
9. BARNETT, S., 'A new formulation of the Liénard–Chipart stability criterion', *Proc. Cambridge Philos. Soc.* **70** (1971).
10. BÔCHER, M., *Introduction to higher algebra*, Dover, New York (1964) (reprint of 1907 edn.).

11. DUFFIN, R. J., 'Algorithms for classical stability problems', *SIAM Review*, **11**, 196–213 (1969).
12. EPSTEIN, M. P., 'The use of resultants to locate extreme values of polynomials', *SIAM J. Appl. Math.* **16**, 62–70 (1968).
13. FULLER, A. T., 'Stability criteria for linear systems and realizability criteria for RC networks', *Proc. Cambridge Philos. Soc.* **53**, 878–896 (1957).
14. FRYER, W. D., 'Applications of Routh's algorithm to network-theory problems', *IRE Trans. Circuit Theory*, **CT-6**, 144–149 (1959).
15. GANTMACHER, F. R., *Theory of matrices*, vols. I, II, Chelsea Publishing Company, New York (1960).
16. HOUSEHOLDER, A. S., 'Bigradients and the problem of Routh and Hurwitz', *SIAM Review*, **10**, 56–66 (1968).
17. HOUSEHOLDER, A. S., 'Bigradients, Hankel determinants, and the Padé table', in *Constructive aspects of the fundamental theorem of algebra*, Ed. B. Dejon and P. Henrici, Wiley-Interscience, New York (1969).
18. HOUSEHOLDER, A. S., 'Bezoutiants, elimination and localization', *SIAM Review*, **12**, 73–78 (1970).
19. JURY, E. J., 'On the roots of a real polynomial inside the unit circle and a stability criterion for linear discrete systems', *Proc. 2nd IFAC Congress (Theory)*, 142–153, Butterworth, London (1964).
20. KALMAN, R. E., 'Mathematical description of linear dynamical systems', *SIAM J. Control*, **1**, 152–192 (1963).
21. KALMAN, R. E., 'Some computational problems and methods related to invariant factors and control theory', in *Computational problems in abstract algebra*, Ed. J. Leech, 393–398, Pergamon, London (1970).
22. LAIDACKER, M. A., 'Another theorem relating Sylvester's matrix and the greatest common divisor', *Math. Mag.* **42**, 126–128 (1969).
23. LEHNIGK, S. H., *Stability theorems for linear motions with an introduction to Liapunov's direct method*, Prentice-Hall, Englewood Cliffs, New Jersey (1966).
24. LEHNIGK, S. H. 'Liapunov's direct method and the number of zeros with positive real parts of a polynomial with constant complex coefficients', *SIAM J. Control*, **5**, 234–244 (1967).
25. MACDUFFEE, C. C., 'Some applications of matrices in the theory of equations', *Amer. Math. Monthly*, **57**, 154–161 (1950).
26. MARDEN, M., *Geometry of polynomials*, American Math. Society, Providence, Rhode Island (1966).

27. Mishina, A. P. and Proskuryakov, I. V., *Higher Algebra*, Pergamon, London (1965).
28. Muir, T., *The theory of determinants in historical order of development*, vols. I–IV, Dover, New York, 1960 (reprint).
29. Muir, T., *A treatise on the theory of determinants*, Dover, New York (1960) (reprint of 1933 Edn.).
30. Parks, P. C., 'Analytic methods for investigating stability—linear and non-linear systems. A survey', *Proc. I. Mech. E.* **178,** Pt 3M (1963–1964).
31. Parks, P. C., 'Liapunov and the Schur–Cohn stability criterion', *IEEE Trans. Aut. Control*, **AC-9,** 121 (1964).
32. Parks, P. C., 'Hermite-Hurwitz and Hermite–Bilharz links using matrix multiplication', *Electronics Letters*, **5,** 55–57 (1969).
33. Rosenbrock, H. H., *State-space and multivariable theory*, Nelson, London (1970).
34. Salmon, G., *Lessons introductory to the modern higher algebra*, Chelsea Publishing Company, New York (1964) (reprint of 1885 Edn.).
35. Turnbull, H. W., *Theory of equations*, 5th Edn., Oliver and Boyd, Edinburgh (1952).
36. Van Vleck, E. B., *On the determination of a series of Sturm's functions by the calculation of a single determinant*, Ann. Math., Second Series, **1,** 1–13 (1899–1900).
37. Wall, H. S., *Analytic theory of continued fractions*, Van Nostrand, New York (1948).

3

Rational Matrices

3.1 McMillan form of a rational matrix

A *rational* $m \times l$ matrix $G(\lambda) = [g_{ij}(\lambda)]$ is one with elements which are rational functions of λ with real or complex coefficients, so we can write

$$g_{ij}(\lambda) = x_{ij}(\lambda)/y_{ij}(\lambda),$$

where $x_{ij}(\lambda)$ and $y_{ij}(\lambda)$ are polynomials. Let $g(\lambda)$ be the monic least common multiple of all the $y_{ij}(\lambda)$. Then $g(\lambda)G(\lambda)$ is a polynomial matrix, having Smith form (Theorem 1.5)

$$S(\lambda) = \text{diag}\,[s_1(\lambda), s_2(\lambda), \ldots, s_R(\lambda), 0, \ldots, 0], \qquad (3.1.1)$$

where $s_1(\lambda), \ldots, s_R(\lambda)$ are the invariant factors of $g(\lambda)G(\lambda)$ and $R = \text{rank}\,\{g(\lambda)G(\lambda)\}$. The specific form of the matrix in (3.1.1) will depend upon the relative magnitudes of m, l and R. Using Theorem 1.3 we can write

$$P_1(\lambda)g(\lambda)G(\lambda)Q_1(\lambda) = S(\lambda),$$

where $P_1(\lambda)$ and $Q_1(\lambda)$ are invertible polynomial matrices, so that we have

$$P_1(\lambda)G(\lambda)Q_1(\lambda) = S(\lambda)/g(\lambda).$$

Cancellation of any common factors in the matrix $S(\lambda)/g(\lambda)$ leads to the following result:

THEOREM 3.1 [16] Every rational matrix $G(\lambda)$ can be expressed as

$$G(\lambda) = P(\lambda)M(\lambda)Q(\lambda), \qquad (3.1.2)$$

where $P(\lambda)$ and $Q(\lambda)$ are invertible polynomial matrices of appropriate dimensions, and the matrix

$$M(\lambda) = \begin{bmatrix} \varepsilon_1(\lambda)/\psi_1(\lambda) & & & & & & & & & \\ & \cdot & & & & & & & & \\ & & \cdot & & & & & & & \\ & & & \cdot & & & & & & \\ & & & & \varepsilon_r(\lambda)/\psi_r(\lambda) & & & & & \\ & & & & & \varepsilon_{r+1}(\lambda) & & & & \\ & & & & & & \cdot & & & \\ & & & & & & & \cdot & & \\ & & & & & & & & \varepsilon_R(\lambda) & \\ & & & & & & & & & 0 \\ & & & & & & & & & & \cdot \\ & & & & & & & & & & & 0 \end{bmatrix} \quad (3.1.3)$$

is the *McMillan* (or *Smith–McMillan*) form of $G(\lambda)$, where

(i) $\varepsilon_i(\lambda)$ and $\psi_i(\lambda)$ are relatively prime monic polynomials
(ii) $\varepsilon_i(\lambda)/\psi_i(\lambda) = s_i(\lambda)/g(\lambda)$
(iii) $\varepsilon_i(\lambda)$ is a factor of $\varepsilon_{i+1}(\lambda)$, $i = 1, 2, \ldots, R-1$, and $\varepsilon_1(\lambda) = s_1(\lambda)$
(iv) $\psi_i(\lambda)$ is a factor of $\psi_{i-1}(\lambda)$, $i = 1, 2, \ldots, r$, and $\psi_1(\lambda) = g(\lambda)$.

The properties (i)–(iv) easily follow from the fact (Theorem 1.5) that $s_i(\lambda)$ is a factor of $s_{i+1}(\lambda)$, $i = 1, 2, \ldots, R-1$.

The integer $r \leqslant R$ is termed the *subrank* of $G(\lambda)$. Clearly the polynomials $\varepsilon_i(\lambda)$ and $\psi_i(\lambda)$ are uniquely determined by $G(\lambda)$.

For example, if

$$G(\lambda) = \begin{bmatrix} \dfrac{1}{(\lambda-1)^2} & \dfrac{1}{(\lambda-1)(\lambda+3)} \\ \dfrac{-6}{(\lambda-1)(\lambda+3)^2} & \dfrac{\lambda-2}{(\lambda+3)^2} \end{bmatrix},$$

then

$$g(\lambda) = (\lambda-1)^2(\lambda+3)^2,$$

$$g(\lambda)G(\lambda) = \begin{bmatrix} (\lambda+3)^2 & (\lambda-1)(\lambda+3) \\ -6(\lambda-1) & (\lambda-2)(\lambda-1)^2 \end{bmatrix},$$

$$S(\lambda) = \begin{bmatrix} 1 & 0 \\ 0 & \lambda(\lambda+1)(\lambda+3)(\lambda-1)^2 \end{bmatrix},$$

and

$$M(\lambda) = \begin{bmatrix} \dfrac{1}{(\lambda-1)^2(\lambda+3)^2} & 0 \\ 0 & \dfrac{\lambda(\lambda+1)}{\lambda+3} \end{bmatrix}.$$

A rational matrix is called *proper* if $\delta x_{ij} < \delta y_{ij}$, all i, j, so that $G(\lambda) \to 0$ as $\lambda \to \infty$ and is said to be *finite at infinity* if $G(\lambda)$ tends to a matrix with finite elements as $\lambda \to \infty$. The example just given illustrates that a proper rational matrix will not necessarily have a proper McMillan form. The set of all the factors $(\lambda - \alpha)^\beta$ of the $\psi_j(\lambda)$ for a proper rational matrix have been termed the *elementary divisors* of $G(\lambda)$ [13]. We also say there is a finite *pole* of $G(\lambda)$ at $\lambda = \alpha$ if any element of $G(\lambda)$ has a pole at $\lambda = \alpha$ (i.e. if any $y_{ij}(\lambda)$ has a factor $(\lambda - \alpha)^\beta$). From Theorem 3.1 it follows that the finite poles of $G(\lambda)$ are the zeros of the polynomials $\psi_j(\lambda)$ in its McMillan form (3.1.3). Similarly, the zeros of the $\varepsilon_i(\lambda)$ are called the *zeros* of $G(\lambda)$.

As well as playing a fundamental role in the theory of linear control systems, as we shall see in the remainder of this chapter, rational matrices are also of considerable importance in certain areas of network theory [17].

EXERCISE

3.1.1 What can be said, if anything, about the effect on the $\varepsilon_i(\lambda)$ and $\psi_j(\lambda)$ of adding a polynomial matrix to $G(\lambda)$?

3.2 Realization theory

Consider the linear system

$$\left.\begin{aligned} \dot{x} &= Ax + Bu \\ y &= Cx \end{aligned}\right\}, \tag{3.2.1}$$

where A is $n \times n$, B is $n \times l$ and C is $m \times n$. The equations (3.2.1) are obtained from (1.4.1) and (1.4.3) with $D(\lambda) \equiv 0$, so from (1.4.4) the transfer function matrix is the $m \times l$ rational matrix

$$G(\lambda) = C(\lambda I_n - A)^{-1}B. \tag{3.2.2}$$

The problem we consider in this section is a converse one: given a proper rational $m \times l$ matrix $G(\lambda)$, determine matrices A, B and C such that (3.2.2) holds. Such a set $\{A, B, C\}$ is termed a *realization* of $G(\lambda)$. If $G(\lambda)$ is not proper, then by straightforward division it can be expressed in the form $G_0(\lambda) + D(\lambda)$, where $G_0(\lambda)$ is proper and $D(\lambda)$ is a polynomial matrix. The modified form of the realization given in (1.4.4) then applies, so that

$$G(\lambda) = C(\lambda I_n - A)^{-1}B + D(\lambda). \tag{3.2.3}$$

In this section we shall therefore assume from now on that $G(\lambda)$ is proper, and in general there will exist infinitely many suitable realizations (3.2.2). Amongst these some will incorporate a matrix A of least dimensions; such a realization is termed *minimal*, and its *dimension* is the order of the minimal A. We shall call this dimension the *degree*

$\delta[G(\lambda)]$ of the rational matrix $G(\lambda)$. Several other definitions of degree have been given and a thorough treatment of relationships between them, including extension to the case of rational matrices which are not proper, may be found in [13]. The degree of a rational matrix has the following properties (see [13] for proofs).

THEOREM 3.2

(i) $\delta[G] \geqslant 0$.
(ii) $\delta[G] = 0$ implies that $G(\lambda)$ is independent of λ.
(iii) If $G^{-1}(\lambda)$ exists, $\delta[G^{-1}] = \delta[G]$.
(iv) If $G_1(\lambda)$ and $G_2(\lambda)$ are two rational $m \times l$ matrices,

$$\delta[G_1] + \delta[G_2] \geqslant \delta[G_1 + G_2],$$

with equality holding if $G_1(\lambda)$ and $G_2(\lambda)$ have no pole (finite or infinite) in common.

(v) If $G_1(\lambda)$ and $G_2(\lambda)$ are conformable for multiplication,

$$\delta[G_1 G_2] \leqslant \delta[G_1] + \delta[G_2].$$

These properties agree with those of the degree of a scalar rational function of λ (see Exercise 3.2.1).

The next theorem illustrates the central role played by the McMillan form.

THEOREM 3.3 [13] The dimension of a minimal realization of a proper rational matrix $G(\lambda)$ is $\sum_{i=1}^{r} \delta\{\psi_i(\lambda)\}$, where $\psi_i(\lambda)$ are given in Theorem 3.1.

In fact, this is McMillan's definition of degree [16].

We now give a relationship between any two minimal realizations.

THEOREM 3.4 [12] If $\{A_1, B_1, C_1\}$ and $\{A_2, B_2, C_2\}$ are two minimal realizations of $G(\lambda)$, then there exists a nonsingular matrix T such that

$$\left.\begin{aligned} A_2 &= TA_1T^{-1} \\ B_2 &= TB_1 \\ C_2 &= C_1T^{-1} \end{aligned}\right\}. \tag{3.2.4}$$

Conversely, if $\{A_1, B_1, C_1\}$ is a minimal realization and T is nonsingular, $\{A_2, B_2, C_2\}$ as given by (3.2.4) is also minimal.

Also of great importance is:

THEOREM 3.5 [12] A realization is minimal if and only if it is controllable and observable†.

† *See Appendix* 2.

The elementary divisors of $G(\lambda)$, defined in the previous section, are directly related to a minimal realization:

THEOREM 3.6 [13] In any minimal realization of $G(\lambda)$ the elementary divisors of the polynomial matrix $\lambda I_n - A$ are the same as the elementary divisors of the rational matrix $G(\lambda)$.

Notice that in this theorem the term 'elementary divisors' is used in two different senses. The proof is based on a procedure for constructing minimal realizations, which involves representing $G(\lambda)$ in partial fractions as $\sum_i G_i(\lambda)$, where each element of the rational matrix $G_i(\lambda)$ has a single pole (of various orders) at λ_i, and the λ_i are all distinct. Kalman then shows how to obtain a minimal realization of each $G_i(\lambda)$ using its McMillan form, and the realization for $G(\lambda)$ is given by

THEOREM 3.7 Let $\{A_i, B_i, C_i\}$ be a minimal realization of $G_i(\lambda)$, $i = 1, 2, \ldots, q$. Then

$$G(\lambda) = \sum_{i=1}^{q} G_i(\lambda)$$

has a realization $\{A, B, C\}$, where

$$A = \text{diag}\,[A_1, A_2, \ldots, A_q]$$

$$B = \begin{bmatrix} B_1 \\ B_2 \\ \cdot \\ \cdot \\ \cdot \\ B_q \end{bmatrix}$$

$$C = [C_1, C_2, \ldots, C_q],$$

which is also minimal provided A_i and A_j have no characteristic roots in common $(i, j = 1, 2, \ldots, q)$. The realization $\{A, B, C\}$ is termed the *direct sum* of $\{A_i, B_i, C_i\}$.

Instead of going into details of this method, we outline a rather simpler approach given by Kalman in 1966 [14]. Let the columns of $P(\lambda)$ and rows of $Q(\lambda)$ in (3.1.2) be $p_1(\lambda), \ldots, p_m(\lambda)$ and $q_1(\lambda), \ldots, q_l(\lambda)$ respectively. Then

$$G(\lambda) = \sum_{k=1}^{R} p_k q_k \varepsilon_k / \psi_k \qquad (\text{with } \psi_k = 1, k > r)$$

$$= \sum_{k=1}^{R} G_k(\lambda) \tag{3.2.5}$$

where

$$G_k(\lambda) = p_k q_k \varepsilon_k / \psi_k. \tag{3.2.6}$$

For $r < k \leqslant R$, $G_k(\lambda)$ is a polynomial matrix; for $k \leqslant r$, $G_k(\lambda)$ is a rational matrix but will not necessarily be proper. By division in (3.2.6), we can write (3.2.5) as $\sum_{k=1}^{R} \{H_k(\lambda) + J_k(\lambda)\}$, where the $H_k(\lambda)$ $(k \leqslant r)$ are now proper and the $J_k(\lambda)$ are polynomial matrices. Since $G(\lambda)$ is proper it follows that $\sum J_k(\lambda) = 0$, so we need therefore only consider the problem of obtaining a realization for $\sum_{k=1}^{r} H_k(\lambda)$. Let A_k be the companion matrix in the form (2.2.1) of

$$\psi_k(\lambda) = \lambda^x + \tau_1 \lambda^{x-1} + \cdots + \tau_x$$

and let

$$\alpha_k(\lambda) = \begin{bmatrix} 1 \\ \lambda \\ \lambda^2 \\ \cdot \\ \cdot \\ \cdot \\ \lambda^{x-1} \end{bmatrix}, \quad \beta_k(\lambda) = \begin{bmatrix} \lambda^{x-1} + \tau_1 \lambda^{x-2} + \cdots + \tau_{x-1} \\ \lambda^{x-2} + \tau_1 \lambda^{x-3} + \cdots + \tau_{x-2} \\ \cdot \\ \cdot \\ \cdot \\ \lambda + \tau_1 \\ 1 \end{bmatrix}.$$

For any polynomial matrix (or vector) $X(\lambda)$ denote by $[X(\lambda)]_M$ the remainder after division of every element of $X(\lambda)$ by $\psi_k(\lambda)$. It is not difficult to show (see Exercise 3.2.7) that

$$\psi_k(\lambda)(\lambda I_x - A_k)^{-1} = [\alpha_k(\lambda)\beta_k^T(\lambda)]_M. \tag{3.2.7}$$

It then follows, using (3.2.7), that a realization of $H_k(\lambda)$ is $\{A_k, B_k, C_k\}$ where B_k and C_k are uniquely determined by

$$[p_k(\lambda)]_M = C_k \alpha_k(\lambda) \tag{3.2.8}$$

$$[\varepsilon_k(\lambda)\, q_k(\lambda)]_M = \beta_k^T(\lambda)\, B_k. \tag{3.2.9}$$

The dimension of this realization of $H_k(\lambda)$ is of course $x = \delta\{\psi_k(\lambda)\}$. By Theorem 3.7 the direct sum of the realizations of $H_k(\lambda)$, $k = 1, 2, \ldots, r$ is a realization of $\sum_{k=1}^{r} H_k(\lambda)$ and has dimension $\sum_{k=1}^{r} \delta\{\psi_k(\lambda)\}$, so that by Theorem 3.3 it is a minimal realization.

Consider as an example

$$G(\lambda) = \frac{1}{g(\lambda)} \begin{bmatrix} \lambda^2 + 6 & \lambda^2 + \lambda + 4 \\ 2\lambda^2 - 7\lambda - 2 & \lambda^2 - 5\lambda - 2 \end{bmatrix} \tag{3.2.10}$$

where $g(\lambda) = \lambda^3 + 2\lambda^2 - \lambda - 2 = (\lambda - 1)(\lambda + 1)(\lambda + 2)$.

The McMillan form of $G(\lambda)$ is

$$\begin{bmatrix} 1/g(\lambda) & 0 \\ 0 & 2 - \lambda \end{bmatrix},$$

so $\psi_1(\lambda) = g(\lambda)$ and $r = 1$. A suitable pair of matrices in (3.1.2) is

$$P(\lambda) = \begin{bmatrix} 1 & 0 \\ \dfrac{\lambda^3 + 2\lambda^2 - 11\lambda - 2}{10} & 1 \end{bmatrix}, \quad Q(\lambda) = \begin{bmatrix} \lambda^2 + 6 & \lambda^2 + \lambda + 4 \\ \dfrac{\lambda + 2}{10} & \dfrac{\lambda + 3}{10} \end{bmatrix}.$$

Also

$$A_1 = \begin{bmatrix} 0 & 1 & 0 \\ 0 & 0 & 1 \\ 2 & 1 & -2 \end{bmatrix}, \tag{3.2.11}$$

and (3.2.8) and (3.2.9) give

$$C_1 = \begin{bmatrix} 1 & 0 & 0 \\ 0 & -1 & 0 \end{bmatrix}, \quad B_1 = \begin{bmatrix} 1 & 1 \\ -2 & -1 \\ 11 & 7 \end{bmatrix}. \tag{3.2.12}$$

It is easy to verify that $C_1(\lambda I_3 - A_1)^{-1}B$ does indeed equal $G(\lambda)$. The degree of $G(\lambda)$ is thus equal to 3, the order of A_1.

Notice that one difficulty in applying the method lies in finding the matrices $P(\lambda)$ and $Q(\lambda)$ (see the remarks following Theorem 1.5). Of course, these matrices are not unique, but any minimal realizations obtained will be related through Theorem 3.4. In fact Kalman [14] treats the same example and obtains two alternative minimal realizations.

In an earlier paper [12] Kalman also gives two other methods for determining minimal realizations. The first, due originally to Gilbert [9] is more straightforward but is applicable only when no element of $G(\lambda)$ has multiple poles:

THEOREM 3.8 Let $G(\lambda)$ have simple poles λ_i, $i = 1, 2, \ldots, p$ and let $K_i = \lim_{\lambda \to \lambda_i} (\lambda - \lambda_i)G(\lambda)$ have rank r_i. Then if

$$K_i = L_i M_i, \tag{3.2.13}$$

where L_i is $m \times r_i$ and M_i is $r_i \times l$, both matrices having rank r_i, a minimal realization of $G(\lambda)$ is

$$A = \operatorname{diag}\,[\lambda_1 I_{r_1}, \lambda_2 I_{r_2}, \ldots, \lambda_p I_{r_p}],$$

$$B = \begin{bmatrix} M_1 \\ M_2 \\ \cdot \\ \cdot \\ \cdot \\ M_p \end{bmatrix},$$

$$C = [L_1, L_2, \ldots, L_p].$$

Incidentally, it was in dealing with the case when $l = m = 1$ that Kalman rediscovered a matrix form of the resultant of two polynomials (see Exercise 2.2.3).

The factorization in (3.2.13) is straightforward to accomplish and a computational procedure is suggested by Kalman [12] in an appendix. A disadvantage of Gilbert's method, however, lies in the need to calculate the λ_i.

Panda and Chen [18] give a method for obtaining a minimal realization which produces a suitable A in Jordan form. Other procedures for finding minimal realizations are described in [10], [11], [15] and the next section. Several authors claim that their methods are efficient, but there is clearly a need for a critical numerical comparison of the various algorithms.

EXERCISES

3.2.1 If $x(\lambda)$ and $y(\lambda)$ are two polynomials, the degree of

$$g(\lambda) = x(\lambda)/y(\lambda)$$

is defined as the sum of the orders of the poles of $g(\lambda)$ (including the pole at infinity). Show that

$$\delta[g(\lambda)] = \max\,[\delta\{x(\lambda)\}, \delta\{y(\lambda)\}]$$

and verify that the properties (i)–(v) of Theorem 3.2 hold.

3.2.2 Show that in equation (3.2.2) $G(\lambda)$ will have a pole where λ is equal to a characteristic root of A.

3.2.3 If A is any $n \times n$ matrix, Leverrier's algorithm [8, p. 87] states that

$$(\lambda I_n - A)^{-1} = \text{Adj}\,(\lambda I_n - A)/\det\,(\lambda I_n - A)$$

where

$$\det\,(\lambda I_n - A) = \lambda^n + p_1\lambda^{n-1} + \cdots + p_n,$$
$$\text{Adj}\,(\lambda I_n - A) = \lambda^{n-1}I_n + \lambda^{n-2}A^{(1)} + \cdots + A^{(n-1)},$$

and

$$A^{(1)} = A + p_1 I_n,\ A^{(k)} = AA^{(k-1)} + p_k I_n,\ k = 2, 3, \ldots, n-1.$$

Hence, or otherwise, show that if A is in the form (2.2.1) the last column of Adj $(\lambda I_n - A)$ is $[1, \lambda, \ldots, \lambda^{n-1}]^T$.

3.2.4 [12] Let $l = m = 1$ and

$$G(\lambda) = \frac{\beta_1\lambda^{n-1} + \beta_2\lambda^{n-2} + \cdots + \beta_n}{\lambda^n + a_1\lambda^{n-1} + \cdots + a_n}$$

where the two polynomials are relatively prime. Verify (using Exercise 3.2.3) that a minimal realization is given by

$$B = [0, 0, \ldots, 0, 1]^T, \quad C = [\beta_n, \beta_{n-1}, \ldots, \beta_1]$$

and A the companion matrix of the denominator in the form (2.2.1).

3.2.5 [12] When $l = 1$, $m > 1$, let

$$G(\lambda) = \frac{1}{a(\lambda)} \begin{bmatrix} \beta_{11}\lambda^{n-1} + \cdots + \beta_{1n} \\ \vdots \\ \beta_{m1}\lambda^{n-1} + \cdots + \beta_{mn} \end{bmatrix},$$

where $a(\lambda) = \lambda^n + a_1\lambda^{n-1} + \cdots + a_n$ is the least common denominator of the elements of $G(\lambda)$. Verify that a minimal realization is given by A, B as in Exercise 3.2.4 and

$$C = \begin{bmatrix} \beta_{1n} & \cdots & \beta_{11} \\ \vdots & \cdots & \vdots \\ \beta_{mn} & \cdots & \beta_{m1} \end{bmatrix}.$$

3.2.6 [14] If $\{A_1, B_1, C_1\}$ and $\{A_2, B_2, C_2\}$ are realizations of $G_1(\lambda)$ and $G_2(\lambda)$ respectively, show that

$$A = \begin{bmatrix} A_1 & B_1C_2 \\ 0 & A_2 \end{bmatrix}, \quad B = \begin{bmatrix} 0 \\ B_2 \end{bmatrix}, \quad C = [C_1 \quad 0]$$

is a realization of $G_1(\lambda)G_2(\lambda)$.

3.2.7 Verify equation (3.2.7) by multiplying both sides on the right by $\lambda I_x - A_k$.

3.2.8 Show that in (3.2.11) and (3.2.12), $[A_1, B_1]$ is controllable and $[A_1, C_1]$ observable.

3.2.9 Use Theorem 3.8 to obtain a minimal realization for the matrix in (3.2.10). Also, obtain the matrix T in Theorem 3.4 connecting your result with (3.2.11) and (3.2.12).

3.2.10 Obtain the McMillan form of

$$G(\lambda) = \frac{1}{g(\lambda)} \begin{bmatrix} \lambda + 2 & 2(\lambda + 2) \\ -1 & \lambda + 1 \end{bmatrix}$$

where $g(\lambda) = \lambda^2 + 3\lambda + 2$, and hence determine the degree of $G(\lambda)$. Verify by finding $\delta[G(\lambda)]$ using Theorem 3.8.

Notice the fallacy of supposing that since $\delta\{g(\lambda)\} = 2$, then $\delta[G(\lambda)] = 2$.

3.3 Rosenbrock's theory of linear systems

We have already given some details (Chapter 1, Section 4) of the substantial and distinctive contribution made recently by Rosenbrock to the theory of linear systems. Because of the importance of this work in clarifying, extending and unifying many concepts and problems in linear multivariable control theory, we devote a separate section to outlining briefly some of the results which directly involve rational matrices. For a full understanding Rosenbrock's complete account [27] is of course essential; in particular, we do not give any proofs of theorems, etc., as these are developed by Rosenbrock in an elegant and integrated manner in the course of his book.

It will probably be most helpful to the reader if we begin by reconsidering the minimal realization problem of the previous section. Using the terminology introduced in Section 1.4, this problem becomes: given a rational $m \times l$ matrix $G(\lambda)$, find a corresponding *first form* (or *state space form*) *system matrix* (see (1.4.4) and (1.4.5))

$$P(\lambda) = \begin{bmatrix} \lambda I_n - A & B \\ -C & D(\lambda) \end{bmatrix} \tag{3.3.1}$$

which has least order (that is, which has A of least dimension) such that

$$G(\lambda) = C(\lambda I_n - A)^{-1}B + D(\lambda).$$

Notice that if $G(\lambda)$ is proper then $D(\lambda) = 0$. If H is a nonsingular constant matrix then the transformation

$$\begin{bmatrix} H^{-1} & 0 \\ 0 & I_m \end{bmatrix} \begin{bmatrix} \lambda I_n - A & B \\ -C & D(\lambda) \end{bmatrix} \begin{bmatrix} H & 0 \\ 0 & I_l \end{bmatrix}$$

is called *system similarity*. We can now deduce Theorem 3.5, giving a necessary and sufficient condition for a realization to be minimal, from Theorems 1.15 and 1.22, and can rewrite Theorem 3.4 as

THEOREM 3.9 Two system matrices in the form (3.3.1) having least order correspond to the same $G(\lambda)$ if and only if they are system similar.

A straightforward way of obtaining a realization of $G(\lambda)$ is as follows [21, 27]. If $G(\lambda)$ is not proper we write as before

$$G(\lambda) = G_0(\lambda) + D(\lambda), \tag{3.3.2}$$

where $G_0(\lambda)$ is proper and $D(\lambda)$ is a polynomial matrix. Let $g_i(\lambda)$ be the monic least common denominator of row i of $G_0(\lambda)$ and write

$$G_0(\lambda) = \begin{bmatrix} u_1(\lambda)/g_1(\lambda) \\ \cdot \\ \cdot \\ \cdot \\ u_m(\lambda)/g_m(\lambda) \end{bmatrix}$$

with

$$g_i(\lambda) = \sum_{k=0}^{p(i)} g_{ik}\lambda^k \qquad (g_{ip(i)} = 1),$$

$$u_i(\lambda) = \sum_{k=0}^{p(i)-1} u_{ik}\lambda^k. \qquad (u_i(\lambda) \text{ and } u_{ik} \text{ are row } l\text{-vectors.})$$

Then a realization of $G_0(\lambda)$ is

$$\left.\begin{aligned} A &= \operatorname{diag}[A_1, A_2, \ldots, A_m], \\ B &= \begin{bmatrix} B_1 \\ B_2 \\ \cdot \\ \cdot \\ \cdot \\ B_m \end{bmatrix}, \\ C &= [C_1, C_2, \ldots, C_m] \end{aligned}\right\} \tag{3.3.3}$$

where A_i is the companion matrix of $g_i(\lambda)$ in the form

$$\left.\begin{aligned} A_i &= \begin{bmatrix} 0 & 0 & \ldots & 0 & -g_{i0} \\ 1 & 0 & \ldots & 0 & -g_{i1} \\ 0 & 1 & \ldots & 0 & -g_{i2} \\ \cdot & \cdot & \ldots & \cdot & \cdot \\ \cdot & \cdot & \ldots & 0 & \cdot \\ 0 & 0 & \ldots & 1 & -g_{i,p(i)-1} \end{bmatrix} \\ B_i &= \begin{bmatrix} u_{i0} \\ u_{i1} \\ \cdot \\ \cdot \\ \cdot \\ u_{i,p(i)-1} \end{bmatrix} \\ \text{and} \qquad C_i &= \begin{bmatrix} 0 & 0 & \ldots & 0 & \\ \cdot & \cdot & \ldots & \cdot & e_i \\ \cdot & \cdot & \ldots & \cdot & \\ 0 & 0 & \ldots & 0 & \end{bmatrix} \end{aligned}\right\} \tag{3.3.4}$$

where e_i is the ith column of I_m.

For example, with $G(\lambda)$ as in (3.2.10) we have $g_1(\lambda) = g_2(\lambda) = g(\lambda)$, $p_1 = p_2 = 3$, so (3.3.3) and (3.3.4) give

$$A = \begin{bmatrix} 0 & 0 & 2 & 0 & 0 & 0 \\ 1 & 0 & 1 & 0 & 0 & 0 \\ 0 & 1 & -2 & 0 & 0 & 0 \\ 0 & 0 & 0 & 0 & 0 & 2 \\ 0 & 0 & 0 & 1 & 0 & 1 \\ 0 & 0 & 0 & 0 & 1 & -2 \end{bmatrix},$$

$$B = \begin{bmatrix} 6 & 4 \\ 0 & 1 \\ 1 & 1 \\ -2 & -2 \\ -7 & -5 \\ 2 & 1 \end{bmatrix},$$

$$C = \begin{bmatrix} 0 & 0 & 1 & 0 & 0 & 0 \\ 0 & 0 & 0 & 0 & 0 & 1 \end{bmatrix}.$$

Clearly this procedure will not in general give a minimal realization. However it is easy to verify that $\lambda I - A^T$ and C^T in (3.3.3) are relatively prime, so by virtue of Theorem 1.15 the realization (3.3.3) can be reduced if and only if $\lambda I - A$ and B are not relatively prime. Rosenbrock gives an algorithm for carrying out this reduction, and claims that this produces a minimal realization more easily than the methods referred to in Section 2.

We now turn to the *second form polynomial system matrix* (see equation (1.4.8))

$$P(\lambda) = \begin{bmatrix} \overset{r}{T(\lambda)} & \overset{l}{U(\lambda)} \\ -V(\lambda) & W(\lambda) \end{bmatrix}\begin{matrix} r \\ m \end{matrix}, \tag{3.3.5}$$

with $r \geqslant n$ and $G(\lambda) = V(\lambda)T^{-1}(\lambda)U(\lambda) + W(\lambda)$. Clearly (3.3.1) is a special case of (3.3.5), and we recall the definition given in Section 1.4 that the *order* n of the system is $\delta\{\det T(\lambda)\}$. A transformation which preserves the transfer function matrix (see Exercise 1.4.1) and system order is *strict system equivalence*:

$$\begin{bmatrix} M(\lambda) & 0 \\ X(\lambda) & I_m \end{bmatrix}\begin{bmatrix} T(\lambda) & U(\lambda) \\ -V(\lambda) & W(\lambda) \end{bmatrix}\begin{bmatrix} N(\lambda) & Y(\lambda) \\ 0 & I_l \end{bmatrix},$$

where $M(\lambda)$ and $N(\lambda)$ are invertible polynomial $r \times r$ matrices. Notice that this is less general than the usual definition of equivalence for polynomial matrices. A connection between the two forms of system matrix is provided by

THEOREM 3.10 Any system matrix in second form having order n can be transformed by strict equivalence into the form

$$\begin{bmatrix} I_{r-n} & 0 \\ 0 & P_f(\lambda) \end{bmatrix},$$

where $P_f(\lambda)$ is a system matrix in first form.

System similarity is obviously a special case of strict system equivalence, but the following relationship can be established.

THEOREM 3.11 Two system matrices in first form are system similar if and only if they are strictly system equivalent.

The minimal realization problem is now to find a system matrix in second form, having least order $\nu(G)$, corresponding to a given $G(\lambda)$. In a given realization (3.3.5), unless both $[T(\lambda), U(\lambda)]$ and $[T^T(\lambda), V^T(\lambda)]$ are relatively prime, then by virtue of Theorem 1.15 there will exist a system matrix giving rise to the same $G(\lambda)$ but having lower order. In this case Rosenbrock gives a procedure for obtaining a system matrix $P_1(\lambda)$ in which $[T_1(\lambda), U_1(\lambda)]$ and $[T_1^T(\lambda), V_1^T(\lambda)]$ *are* both relatively prime, so that this constitutes a way of obtaining a minimal realization of $G(\lambda)$. Instead of giving details of this method, which while straightforward is somewhat lengthy, we give solutions to the simpler problem of computation of $\nu(G)$. It is naturally useful to be able to calculate $\nu(G)$ without actually having to find a minimal realization.

THEOREM 3.12 [27] The least order $\nu(G)$ of a rational $m \times l$ matrix $G(\lambda)$ is equal to $\delta\{\phi(\lambda)\}$, where $\phi(\lambda)$ is the monic least common denominator of all the minors of orders $1, 2, \ldots, q$ of $G(\lambda)$ ($q = \min(l, m)$).

This theorem is useful when l and m are small, but otherwise the following involves less effort.

THEOREM 3.13 [20] Write

$$G(\lambda) = H(\lambda)/g(\lambda) + D(\lambda),$$

where $g(\lambda)$ is the monic least common denominator of the elements of $G(\lambda)$, so that $H(\lambda)/g(\lambda)$ is proper and $D(\lambda)$ is a polynomial matrix. Let

$$g(\lambda) = \lambda^p + g_{p-1}\lambda^{p-1} + \cdots + g_0,$$
$$H(\lambda) = H_{p-1}\lambda^{p-1} + H_{p-2}\lambda^{p-2} + \cdots + H_0,$$

and define the $m \times l$ matrices

$$Q_{10} = 0,\ Q_{1j} = H_{j-1}, \qquad (j = 1, 2, \ldots, p)$$
$$Q_{i0} = 0,\ Q_{ij} = Q_{i-1,j-1} - g_{j-1}Q_{i-1,p}$$
$$(i = 2, 3, \ldots, p;\ j = 1, 2, \ldots, p).$$

Then $\nu(G)$ is equal to the rank of the $mp \times lp$ matrix $[Q_{ij}]$, $i, j = 1, 2, \ldots, p$.

It is interesting to notice that the matrix $[Q_{ij}]$ can be written in the form

$$Q = \begin{bmatrix} \tilde{H} \\ \tilde{H}(\Gamma \otimes I_l) \\ \tilde{H}(\Gamma \otimes I_l)^2 \\ \cdot \\ \cdot \\ \cdot \\ \tilde{H}(\Gamma \otimes I_l)^{p-1} \end{bmatrix} \qquad (3.3.6)$$

where $\tilde{H} = [H_0, H_1, \ldots, H_{p-1}]$ and Γ is the companion matrix of $g(\lambda)$ in the form (2.2.1), namely

$$\Gamma = \begin{bmatrix} 0 & 1 & 0 & & \\ 0 & 0 & 1 & \cdot & \\ \cdot & \cdot & \cdot & \cdot & \\ -g_0 & -g_1 & -g_2 & \cdots & -g_{p-1} \end{bmatrix}.$$

The expression (3.3.6) is clearly a generalization of that in (2.2.21). This is not surprising, since a straightforward interchange of rows and columns establishes that Q is equivalent to

$$H_{p-1} \otimes \Gamma^{p-1} + H_{p-2} \otimes \Gamma^{p-2} + \cdots + H_0 \otimes I_p$$

which is of the same form as the matrix in (1.3.1), as it appears in the proof of Theorem 2.11. Thus by Theorem 1.13,

$$\text{rank } Q = qp - \sum_{i=1}^{q} \delta_i$$

where δ_i is the degree of the g.c.d. of $g(\lambda)$ and the ith invariant factor of $H(\lambda)$, and $q = \min(l, m)$; in particular, if $l = 1$ this reduces to Theorem 2.11.

A similar result, which however involves more computation, is due to Ho and Kalman [11]:

THEOREM 3.14 With notation as in Theorem 3.13, expand $H(\lambda)$ in a Laurent series, convergent for large enough $|\lambda|$,

$$H(\lambda) = G_1\lambda^{-1} + G_2\lambda^{-2} + G_3\lambda^{-3} + \cdots$$

Then $\nu(G)$ is equal to the rank of the $mp \times lp$ matrix

$$\begin{bmatrix} G_1 & G_2 & \ldots & G_p \\ G_2 & G_3 & \ldots & G_{p+1} \\ \cdot & \cdot & \ldots & \cdot \\ G_p & G_{p+1} & \ldots & G_{2p-1} \end{bmatrix}.$$

The reader will, of course, be aware that for system matrices in first form, the least order $\nu(G)$ is identical to the degree $\delta[G]$ of $G(\lambda)$ as defined in Section 2. Care is needed, however, when $G(\lambda)$ is not proper and it turns out that in (3.3.2), $\delta[G(\lambda)] = \nu\{G_0(\lambda)\} + \nu\{D(\lambda^{-1})\}$.

We end this selection from Rosenbrock's work with a few theorems which reveal further aspects of his approach. First, suppose that $P(\lambda)$ in (3.3.5) has $l = m$ and is of least order. Let $S_T(\lambda)$ and $S_P(\lambda)$ denote the Smith forms of $T(\lambda)$ and $P(\lambda)$ respectively. It can be shown that $S_T(\lambda)$ has at most m non-unit entries, and since $P(\lambda)$ has least order (so that $T(\lambda)$ and $U(\lambda)$ are relatively prime) it also follows that in $S_P(\lambda)$ the first r elements on the diagonal are units. We can therefore write

$$S_T(\lambda) = \text{diag}\,[1, 1, \ldots, 1, \psi_m(\lambda), \psi_{m-1}(\lambda), \ldots, \psi_1(\lambda)]$$
$$S_P(\lambda) = \text{diag}\,[1, 1, \ldots, 1, \varepsilon_1(\lambda), \varepsilon_2(\lambda), \ldots, \varepsilon_m(\lambda)],$$

where the usual division properties hold (Theorem 1.5).

THEOREM 3.15 [24]

(i) The matrices $\text{diag}\,[\varepsilon_1, \varepsilon_2, \ldots, \varepsilon_m]$ and $\text{diag}\,[\psi_1, \psi_2, \ldots, \psi_m]$ are relatively prime.

(ii) The McMillan form of $G(\lambda)$ is $\text{diag}\,[\varepsilon_1/\psi_1, \varepsilon_2/\psi_2, \ldots, \varepsilon_m/\psi_m]$.

This unexpectedly simple theorem illustrates the fundamental nature of the second form system matrix. Minor modifications are easily made if $l \neq m$.

Standard forms for system matrices in first and second form are of importance in developing the theory. We give the simplest of these:

THEOREM 3.16 If $P(\lambda)$ in (3.3.5) has least order, then it is strictly system equivalent to *either*

(i)
$$\begin{bmatrix} I_{r-m} & 0 & 0 \\ 0 & T_1(\lambda) & U_1(\lambda) \\ 0 & -I_m & D(\lambda) \end{bmatrix}, \tag{3.3.7}$$

where the polynomial matrices $T_1(\lambda)$ and $U_1(\lambda)$ are unique and relatively prime, and $T_1(\lambda) = [t_{ij}(\lambda)]$ is $m \times m$ lower triangular with t_{jj} monic and $\delta\{t_{jj}(\lambda)\} > \delta\{t_{ij}(\lambda)\}$, $i, j = 1, 2, \ldots, m$. If $t_{ii}(\lambda) = 1$ then $t_{ij}(\lambda) = 0$, all $i \neq j$

or

(ii)
$$\begin{bmatrix} I_{r-l} & 0 & 0 \\ 0 & T_2(\lambda) & I_l \\ 0 & -V_2(\lambda) & D(\lambda) \end{bmatrix}, \tag{3.3.8}$$

where the polynomial matrices $T_2(\lambda)$ and $V_2(\lambda)$ are unique, $T_2^T(\lambda)$ and $V_2^T(\lambda)$ are relatively prime, and $T_2(\lambda) = [t_{ij}(\lambda)]$ is lower triangular, with $t_{ii}(\lambda)$ monic and $\delta\{t_{ii}(\lambda)\} > \delta\{t_{ij}(\lambda)\}$, $i, j = 1, 2, \ldots, l$. If $t_{ij}(\lambda) = 1$ then $t_{ij}(\lambda) = 0$, all $j \neq i$.

The polynomial matrix $D(\lambda)$ is uniquely determined by $G(\lambda)$, the invariant polynomials of $U_1(\lambda) + T_1(\lambda)D(\lambda)$ or of $V_2(\lambda) + D(\lambda)T_2(\lambda)$ are the $\varepsilon_i(\lambda)$ in the McMillan form $M_G(\lambda)$ of $G(\lambda)$, and the invariant polynomials of $T_1(\lambda)$ or $T_2(\lambda)$ are, in reverse order, the $\psi_i(\lambda)$ in $M_G(\lambda)$.

Standard forms for system matrices not having least order may also be obtained.

Finally, other interesting questions concern the product of two transfer function matrices $G_2(\lambda)$ and $G_1(\lambda)$ having dimensions $m \times k$ and $k \times l$ respectively ($m \geqslant k, l \geqslant k$). For example, suppose that $G_1(\lambda)$ has a system matrix of least order of the form (3.3.7), so that

$$G_1(\lambda) = T_1^{-1}(\lambda)U_1(\lambda) + D_1(\lambda),$$

and similarly that

$$G_2(\lambda) = V_2(\lambda)T_2^{-1}(\lambda) + D_2(\lambda)$$

from (3.3.8). Then

THEOREM 3.17 If no zero of $G_1(\lambda)$ is a pole of $G_2(\lambda)$, and vice versa, then $G_2(\lambda)G_1(\lambda)$ has least order system matrix

$$\begin{bmatrix} I_{n_1+n_2-k} & 0 & 0 \\ 0 & T_1(\lambda)T_2(\lambda) & \{U_1(\lambda) + T_1(\lambda)D_1(\lambda)\} \\ 0 & -\{V_2(\lambda) + D_2(\lambda)T_2(\lambda)\} & 0 \end{bmatrix},$$

where $n_1 = \delta\{\det T_1(\lambda)\}$ and $n_2 = \delta\{\det T_2(\lambda)\}$. The zeros (poles) of $G_2(\lambda)G_1(\lambda)$ are those of $G_1(\lambda)$ together with those of $G_2(\lambda)$, and $\delta[G_2G_1] = \delta[G_2] + \delta[G_1]$.

We have only been able to give a brief survey of some of the main features of Rosenbrock's work. Nevertheless we hope the reader will have begun to appreciate the value of this approach and will be attracted to the original sources ([21, 24, 25, 26], in addition to those listed in the references for Chapter 1). There will be found many other results involving interesting areas of matrix theory, and giving further insight into basic aspects of the structure of linear multivariable systems such as, for example, controllability, observability and the pole assignment problem (Theorem 4.8). For other material in this same area [10] is also worth consulting.

EXERCISES

3.3.1 [21] Suppose A, B and C in equation (3.3.1) can be partitioned as shown:

$$A = \overset{\quad p \qquad n-p}{\begin{bmatrix} A_{11} & 0 \\ A_{21} & A_{22} \end{bmatrix}}\begin{matrix} p \\ n-p \end{matrix}, \quad B = \begin{bmatrix} 0 \\ B_{21} \end{bmatrix}\begin{matrix} p \\ n-p \end{matrix}, \quad C = [\overset{p}{C_{11}}, \overset{n-p}{C_{12}}].$$

Show that the transfer function matrix corresponding to the system matrix

$$\begin{bmatrix} \lambda I_{n-p} - A_{22} & B_{21} \\ -C_{12} & D(\lambda) \end{bmatrix}$$

is the same as that corresponding to (3.3.1).

3.3.2 Prove that the last row of Adj $(\lambda I_{p(i)} - A_i)$, where A_i is the companion matrix in (3.3.4), is $[1, \lambda, \ldots, \lambda^{p(i)-1}]$, and hence verify that the realization given by (3.3.3) and (3.3.4) is valid.

3.3.3 Apply Theorems 3.12 and 3.13 to the matrix in equation (3.2.10).

3.3.4 Show that

$$A = \begin{bmatrix} 0 & I_l & \cdot & \cdots & \cdot \\ 0 & 0 & I_l & \cdots & \cdot \\ \cdot & \cdot & \cdot & \cdots & \cdot \\ \cdot & \cdot & \cdot & \cdots & I_l \\ -g_0I_l & -g_1I_l & -g_2I_l & \cdots & -g_{p-1}I_l \end{bmatrix},$$

$$B = \begin{bmatrix} 0 \\ 0 \\ \cdot \\ \cdot \\ \cdot \\ 0 \\ I_l \end{bmatrix},$$

$$C = [H_0, H_1, \ldots, H_{p-1}]$$

is a realization of $H(\lambda)/g(\lambda)$ in Theorem 3.13.

3.3.5 Use Theorem 3.15 to show that if $G(\lambda)$ has $\psi_i(\lambda)$ as the denominator polynomials in its McMillan form, then so does $G(\lambda) + G_2(\lambda)$, where $G_2(\lambda)$ is a polynomial matrix.

3.3.6 In Theorem 3.16, prove the statement concerning the invariant polynomials of $U_1 + T_1D$ and $V_2 + DT_2$ by using Theorem 3.15.

3.3.7 Let $G_1(\lambda)$ and $G_2(\lambda)$ be two rational matrices having the same dimensions, corresponding to second form system matrices $P_1(\lambda)$ and $P_2(\lambda)$ respectively. Show that a system matrix for $G_1(\lambda) + G_2(\lambda)$ is

$$\begin{bmatrix} T_1 & 0 & U_1 \\ 0 & T_2 & U_2 \\ -V_1 & -V_2 & W_1 + W_2 \end{bmatrix}.$$

3.3.8 Use Theorem 3.13 to show that if the polynomial matrix $D(\lambda)$ is given by $D_0\lambda^s + D_1\lambda^{s-1} + \cdots + D_s$, then

$$\nu\{D(\lambda^{-1})\} = \text{rank} \begin{bmatrix} D_0 & D_1 & \cdots & D_{s-1} \\ 0 & D_0 & \cdots & D_{s-2} \\ \cdot & \cdot & \cdots & \cdot \\ 0 & 0 & \cdots & D_0 \end{bmatrix}.$$

3.4 Positive-real matrices and spectral factorization

Of importance in network theory [17] and other applications [1] are *positive-real* $n \times n$ matrices $Z(\lambda)$ defined by:

(i) The elements of $Z(\lambda)$ are analytic for $\text{Re}(\lambda) > 0$
(ii) $\bar{Z}(\lambda) = Z(\bar{\lambda})$ for $\text{Re}(\lambda) > 0$
(iii) $Z^T(\bar{\lambda}) + Z(\lambda)$ is positive semidefinite Hermitian for $\text{Re}(\lambda) > 0$.

Generally the elements of $Z(\lambda)$ may be functions of the complex variable λ, but in this section we shall only consider the case when $Z(\lambda)$ is a rational matrix. Condition (ii) then implies that all the coefficients in the elements of $Z(\lambda)$ must be real—such a matrix is termed *real-rational.* This in turn implies that the matrix in (iii) can be written as $Z^*(\lambda) + Z(\lambda)$, and simple properties of positive-real matrices can therefore be deduced using the Hermitian form $x^*\{Z^*(\lambda) + Z(\lambda)\}x$ (see Exercise 3.4.1).

It can be shown [17] that any positive-real matrix can be written as $Z(\lambda) + \lambda R$, where R is positive semidefinite Hermitian and $Z(\lambda)$ tends to a finite matrix as λ tends to infinity. We shall therefore lose no generality by restricting positive-real matrices to be finite at infinity from now on.

Newcomb [17] gives two tests for positive-real matrices:

THEOREM 3.18 An $n \times n$ real-rational matrix $Z(\lambda)$ is positive-real if and only if

(i) it has no poles in $\mathrm{Re}\,(\lambda) > 0$
(ii) any poles on $\mathrm{Re}\,(\lambda) = 0$ are simple
(iii) for each pole on $\mathrm{Re}\,(\lambda) = 0$, the matrix of residues is positive semidefinite Hermitian
(iv) $Z^T(-i\omega) + Z(i\omega)$ is positive semidefinite Hermitian for all real ω, when $i\omega$ is not a pole.

THEOREM 3.19 An $n \times n$ matrix $Z(\lambda)$ is positive-real if and only if

$$X(\lambda) = [Z(\lambda) + I_n]^{-1}[Z(\lambda) - I_n]$$

exists (i.e. $\det\,[Z(\lambda) + I_n] \not\equiv 0$) and satisfies

(i) $X(\lambda)$ is real-rational
(ii) $X(\lambda)$ has no poles in $\mathrm{Re}\,(\lambda) \geqslant 0$
(iii) $I_n - X^*(i\omega)X(i\omega)$ is positive semidefinite Hermitian for all real ω.

Another criterion for positive-realness has recently been given by Siljak in the course of some interesting work on polynomials and polynomial and rational matrices.

THEOREM 3.20 [28] Let $Z(\lambda) = Y(\lambda)/z(\lambda)$, where $Y(\lambda)$ is a real $n \times n$ polynomial matrix and $z(\lambda)$ is a real polynomial having no factors in common with the elements of $Y(\lambda)$. Further, let $[Z(\lambda) + I_n]^{-1} = F(\lambda)/f(\lambda)$ where $f(\lambda)$ and $F(\lambda)$ are also relatively prime. Then $Z(\lambda)$ is positive-real if and only if

(i) $f(\lambda)$ is a Hurwitz polynomial
(ii) the Hermitian matrix $z(-i\omega)Y(i\omega) + z(i\omega)Y^T(-i\omega)$ is positive semidefinite for all real $\omega \geqslant 0$.

This result is easily applied since Siljak [28, 29] has also given necessary and sufficient conditions in terms of Routh arrays for (ii) to be satisfied.

There are important connections between positive-real matrices and minimal realizations. We assume in the next two theorems that $Z(\lambda)$ is proper (otherwise an expression like (3.2.3) applies).

THEOREM 3.21 Let the positive-real matrix $Z(\lambda)$ have no poles on $\mathrm{Re}\,(\lambda) = 0$. If $\{A, B, C\}$ is a minimal realization for $Z(\lambda)$ then all the characteristic roots of A have negative real parts.

The proof is immediate, by means of Theorem 3.6. The condition on the characteristic roots of A is of course a fundamental one for stability of linear systems (see Theorem 4.1). A converse result also holds:

THEOREM 3.22 If $Z(\lambda) = C(\lambda I_n - A)^{-1}B$ is an $n \times n$ real-rational matrix, then it is positive-real if condition (iv) of Theorem 3.18 holds and the characteristic roots of A all have negative real parts.

This also follows at once, since all the conditions of Theorem 3.18 are satisfied.

In order to continue our development we need some more definitions. A real-rational $n \times n$ matrix satisfying $Y(\lambda) = Y^T(-\lambda)$ is termed *parahermitian* and it is easy to see that $Y(i\omega)$ is Hermitian for real ω. A real-rational $m \times n$ matrix $V(\lambda)$ such that $V^T(-\lambda)V(\lambda) = I_n$, or $V(\lambda)V^T(-\lambda) = I_m$ (or both) is called *paraunitary*. A rational matrix is said to have *rank* r if the order of the largest minor which does not vanish identically is r. Other definitions, and related topics, may be found in the valuable text by Newcomb [17]. Unfortunately, since this is one of very few books to contain material on rational matrices, the reader may find the work somewhat difficult to follow because of the general use of network ideas and terminology.

The next topic we consider arises in both network synthesis [17] and multivariable filtering problems [30]: given a parahermitian $n \times n$ matrix $Y(\lambda)$ which is positive semidefinite when $\lambda = i\omega$, find a rational matrix $W(\lambda)$ such that

$$Y(\lambda) = W^T(-\lambda)W(\lambda). \tag{3.4.1}$$

Because of the way in which it arises in multivariable filtering problems, (3.4.1) is usually called *spectral factorization*. It is often required that $W(\lambda)$ should have no poles in $\text{Re}(\lambda) > 0$. Our first result is one due to Youla [30].

THEOREM 3.23 Let the $n \times n$ parahermitian matrix $Y(\lambda)$ have rank r and let $Y(i\omega)$ be positive semidefinite. Then there exists an $r \times n$ real-rational matrix $W(\lambda)$ satisfying (3.4.1), and

(i) $W(\lambda)$ is analytic and has rank r for $\text{Re}(\lambda) > 0$. Also, if $Y(\lambda)$ is analytic for $\text{Re}(\lambda) \geqslant 0$, so is $W(\lambda)$.

(ii) $W(\lambda)$ is unique up to multiplication on the left by an arbitrary constant $r \times r$ real orthogonal matrix.

Further, $W_1(\lambda) = V(\lambda)W(\lambda)$ is a solution of (3.4.1) for an arbitrary $r \times r$ paraunitary matrix $V(\lambda)$, and if $V(\lambda)$ is analytic in $\text{Re}(\lambda) > 0$, so is $W_1(\lambda)$.

The proof of the existence of $W(\lambda)$ having property (i), whilst lengthy (being based once again on the McMillan form) is constructive, and

Youla gives a numerical example. It is of interest to note that if $U(\lambda)$ is the right inverse of $W(\lambda)$, so that $W(\lambda)U(\lambda) = I_r$, then $U(\lambda)$ is also analytic in Re $(\lambda) > 0$. If this condition on $U(\lambda)$ is relaxed, then $W(\lambda)$ in (3.4.1) can be taken triangular, and Youla gives explicit formulae for its elements in this case. We should mention that Youla's results are not restricted to the case $Y(\lambda)$ real-rational.

Instead of going into details of Youla's methods for obtaining $W(\lambda)$, we shall describe a rather ingenious procedure due to Davis [7] which does not explicitly involve finding the McMillan form. The basis of the idea is successively to multiply $Y(\lambda)$ (assumed to have rank n) on the left and right by $n \times n$ matrices $T_i(-\lambda)$ and $T_i^T(\lambda)$ respectively until the resulting product is a real positive definite symmetric matrix with constant elements. The matrix $W(\lambda)$ in (3.4.1) is square and can be easily expressed in terms of the T_i. For simplicity, and to ensure that $W(\lambda)$ and $W^{-1}(\lambda)$ are analytic in Re $(\lambda) > 0$, each $T_i(\lambda)$ is taken to have a simple form so that $T_i(\lambda)$ and $T_i^{-1}(\lambda)$ are analytic in Re $(\lambda) > 0$. The stages are as follows:

(1) $Y(\lambda)$ is transformed into a polynomial matrix

$$Y_1(\lambda) = T_1(-\lambda)Y(\lambda)T_1^T(\lambda),$$

where $T_1(\lambda) = \text{diag}\,[t_{11}(\lambda), \ldots, t_{nn}(\lambda)]$.

Select those denominator factors $\lambda + a_1, \lambda + a_2, \ldots$ in the ith column of $Y(\lambda)$ which have $a_k > 0$. Then $t_{ii}(\lambda) = (\lambda + a_1)(\lambda + a_2) \cdots$

(2) Matrices are determined such that

$$Y_2(\lambda) = T_k(-\lambda) \cdots T_2(-\lambda)Y_1(\lambda)T_2^T(\lambda) \cdots T_k^T(\lambda)$$

has determinant independent of λ.

For a pair of factors $(\lambda + b_i)(-\lambda + b_i)$, with Re $(b_i) > 0$ in det $Y_1(\lambda)$, choose $T_2(\lambda) = \text{diag}\,[1, 1, \ldots, 1, 1/(\lambda + b_i)]$. This removes the above factors from det $Y_1(\lambda)$, but introduces unwanted denominator terms in the last row and column. Remove these by taking

$$T_3(\lambda) = \begin{bmatrix} 1 & 0 & \cdot & \cdot & 0 \\ 0 & 1 & \cdot & \cdot & 0 \\ \cdot & \cdot & \cdot & \cdot & \cdot \\ \cdot & \cdot & \cdot & \cdot & \cdot \\ \dfrac{k_1}{\lambda + b_i} & \dfrac{k_2}{\lambda + b_i} & \cdots & \dfrac{k_{n-1}}{\lambda + b_i} & 1 \end{bmatrix},$$

where the k_i are determined uniquely by making the numerators of the first $n - 1$ elements in the last row of $T_3(-\lambda)T_2(-\lambda)Y_1(\lambda)T_2^T(\lambda)T_3^T(\lambda)$ vanish for $\lambda = b_i$.

Repeat this process until a matrix with constant determinant is obtained.

(3) Next, reduce to

$$Y_3 = T_l(-\lambda) \cdots T_{k+1}(-\lambda) Y_2(\lambda) T_{k+1}^T(\lambda) \cdots T_l^T(\lambda),$$

which is independent of λ.

Form the matrix $Y_2^{(1)}(\lambda)$, say, of highest degree terms in each element of $Y_2(\lambda)$, and replace by zero any which do not contribute to the term of highest degree in det $Y_2^{(1)}(\lambda)$, to give a matrix $Y_2^{(2)}(\lambda)$, say. Since det $Y_2(\lambda)$ is independent of λ, det $Y_2^{(2)}(\lambda)$ must be zero, and it is therefore possible to find a matrix $T_{k+1}(\lambda)$ which makes $T_{k+1}(-\lambda) Y_2^{(2)}(\lambda) T_{k+1}^T(\lambda)$ have a zero row. $T_{k+1}(-\lambda)$ can be taken polynomial and triangular with unit elements on the principal diagonal, by making zero the row of highest degree in $Y_2^{(2)}(\lambda)$.

This procedure is repeated until a constant matrix is obtained.

(4) The matrix Y_3 is real symmetric positive definite, so a real constant matrix Y_4, triangular for simplicity, can be found such that $Y_3 = Y_4 Y_4^T$.

Hence in (3.4.1), $W^T(\lambda) = T_1^{-1}(\lambda) T_2^{-1}(\lambda) \cdots T_l^{-1}(\lambda) Y_4$.

Riddle and Anderson [19] have suggested ways in which (1) and (2) can be efficiently carried out. The above procedure should become clearer if the reader attempts Exercise 3.4.5.

We now turn to some results of Anderson [3] which provide interesting links between positive-real matrices, spectral factorization, realization theory and the Liapunov matrix equation (see Section 1, Chapter 4). If $Z(\lambda)$ is positive-real and has rank r, then $Y(\lambda) = Z(\lambda) + Z^T(-\lambda)$ satisfies the conditions of Theorem 3.23 (see Exercise 3.4.2), so there exists an $r \times n$ matrix $W_1(\lambda)$ such that

$$Y(\lambda) = Z(\lambda) + Z^T(-\lambda) = W_1^T(-\lambda) W_1(\lambda). \qquad (3.4.2)$$

First, for simplicity, suppose that $Z(\lambda)$ is proper and has no poles in Re $(\lambda) \geqslant 0$. We then have the following relationship between minimal realizations of $Z(\lambda)$ and $W_1(\lambda)$ in (3.4.2).

THEOREM 3.24 If $\{A, B, C\}$ and $\{F, G, H\}$ are minimal realizations of $Z(\lambda)$ and $W_1(\lambda)$ respectively, then A and F are similar.

Proof. It is easy to verify that

$$A_1 = \begin{bmatrix} A & 0 \\ 0 & -A^T \end{bmatrix}, B_1 = \begin{bmatrix} B \\ C^T \end{bmatrix}, C_1 = [C \quad -B^T]$$

is a realization for $Z(\lambda) + Z^T(-\lambda)$. Since $Z(\lambda)$ and $Z^T(-\lambda)$ have no poles in common, we have

$$\delta[Z(\lambda)] + \delta[Z^T(-\lambda)] = 2\delta[Z(\lambda)]$$

by Theorem 3.2 (iv), so $\{A_1, B_1, C_1\}$ is minimal.

Also,

$$\begin{aligned} W_1^T(-\lambda)W_1(\lambda) &= [H(-\lambda I_n - F)^{-1}G]^T[H(\lambda I_n - F)^{-1}G] \\ &= G^T(-\lambda I_n - F^T)^{-1}H^T H(\lambda I_n - F)^{-1}G \\ &= H_1(\lambda I_{2n} - F_1)^{-1}G_1, \end{aligned}$$

where

$$F_1 = \begin{bmatrix} F & 0 \\ H^T H & -F^T \end{bmatrix}, \; G_1 = \begin{bmatrix} G \\ 0 \end{bmatrix}, \; H_1 = [0 \quad -G^T]. \qquad (3.4.3)$$

Because of (i) in Theorem 3.23, $W^T(-\lambda)$ and $W(\lambda)$ satisfy the conditions of Theorem 3.17, so that

$$\begin{aligned} \delta[W_1^T(-\lambda)W_1(\lambda)] &= \delta[W_1^T(-\lambda)] + \delta[W_1(\lambda)] \\ &= 2\delta[W_1(\lambda)], \end{aligned}$$

and $\{F_1, G_1, H_1\}$ is therefore a minimal realization for $W_1^T(-\lambda)W_1(\lambda)$. Since this last matrix has no poles in Re $(\lambda) \geqslant 0$, from Theorem 3.21 we know that all the characteristic roots of F have negative real parts. Therefore, since by Theorem 3.5 $[F, H]$ is observable, there exists (see Theorem 4.3) a unique positive definite symmetric matrix P satisfying the Liapunov matrix equation

$$F^T P + PF = -H^T H. \qquad (3.4.4)$$

Since $W_1(\lambda)$ is real-rational, F, H and P are all real. If we now apply Theorem 3.4 to the realization (3.4.3), with

$$T = \begin{bmatrix} I_n & 0 \\ P & I_n \end{bmatrix},$$

where P is given by (3.4.4), we obtain another minimal realization of $W_1^T(-\lambda)W_1(\lambda)$:

$$F_2 = TF_1T^{-1} = \begin{bmatrix} F & 0 \\ 0 & -F^T \end{bmatrix}, \; G_2 = TG_1, \; H_2 = H_1T^{-1}.$$

Finally, since $\{A_1, B_1, C_1\}$ and $\{F_2, G_2, H_2\}$ are both minimal realizations of $Y(\lambda)$, it follows from Theorem 3.4 that A_1 and F_2 are similar. Hence, since all the characteristic roots of A and F have negative real parts, it follows that A and F are also similar.

It is also straightforward to prove:

THEOREM 3.25 With notation as in Theorem 3.24, there exists a matrix K such that $\{A, B, K\}$ is a minimal realization for $W_1(\lambda)$.

Using the previous two results, Anderson then proves:

THEOREM 3.26 Let $Z(\lambda)$ be a proper real-rational matrix analytic in Re $(\lambda) \geqslant 0$, having a minimal realization $\{A, B, C\}$. Then $Z(\lambda)$ is positive-real if and only if there exist a real positive definite symmetric matrix P and real matrix H such that

$$\left.\begin{aligned} A^T P + PA &= -H^T H \\ PB &= C^T. \end{aligned}\right\} \tag{3.4.5}$$

We now drop the restrictions imposed on $Z(\lambda)$ and state Anderson's theorem for a general positive-real matrix.

THEOREM 3.27 Let $Z(\lambda)$ be real-rational and analytic in $\text{Re}(\lambda) > 0$, with any poles on $\text{Re}(\lambda) = 0$ being simple, and let $Z(\lambda) \to Z_0$ as $\lambda \to \infty$. If $\{A, B, C\}$ is a minimal realization of $Z(\lambda)$, then $Z(\lambda)$ is positive-real if and only if there exist a real positive definite symmetric matrix P and real matrices H, J such that

$$\left.\begin{aligned} A^T P + PA &= -H^T H \\ PB &= C^T - H^T J \\ J^T J &= Z_0 + Z_0^T. \end{aligned}\right\} \tag{3.4.6}$$

Notice that since $Z(\lambda)$ is not proper,

$$Z(\lambda) = C(\lambda I_n - A)^{-1} B + Z_0. \tag{3.4.7}$$

Anderson and Moore [5] have called a real-rational matrix $Z(\lambda)$ *generalized positive-real* if it is finite at infinity and satisfies condition (iv) of Theorem 3.18. They obtain a result which is the analogue of Theorem 3.27: If $Z(\lambda)$ is finite at infinity then it is generalized positive-real if and only if there exist a nonsingular real symmetric P and real matrices H, J satisfying (3.4.6). Furthermore, P is not positive definite if $Z(\lambda)$ is not positive-real.

Anderson has also devised [4] a new method of solving the spectral factorization problem which involves yet another important topic in control theory, the matrix Riccati equation (see Chapter 5). As before, let $Y(\lambda)$ be an $n \times n$ parahermitian matrix with $Y(i\omega)$ positive semi-definite, but assume that $Y(\lambda)$ has no poles on $\text{Re}(\lambda) = 0$. Let $Y(\lambda) \to Y_0$ as $\lambda \to \infty$, where Y_0 is finite, and assume that $\det Y_0 \neq 0$, so that Y_0 is real positive definite symmetric. Denote by Y_1 the unique real matrix $(Y_0)^{1/2}$. Let $Z(\lambda)$ be the positive-real matrix given by (3.4.2), having minimal realization $\{A, B, C\}$ as in Theorem 3.27. Anderson discusses the formation of $Z(\lambda)$ for a given $Y(\lambda)$: we have

$$Y(\lambda) = Y^{(1)}(\lambda)/y(\lambda) + Y_0,$$

where the first term is proper, and $y(\lambda) = \Pi_i(\lambda - \lambda_i)(\lambda + \lambda_i)$ with $\text{Re}(\lambda_i) < 0$. If all the λ_i are distinct we can write

$$Y(\lambda) = \sum_i \frac{Y_i}{\lambda - \lambda_i} - \sum_i \frac{Y_i^T}{\lambda + \lambda_i} + Y_0$$

and hence

$$Z(\lambda) = \sum_i \frac{Y_i}{\lambda - \lambda_i} + \frac{Y_0}{2}.$$

A modified expression can be produced if the λ_i are not all distinct. Then:

THEOREM 3.28 A solution of (3.4.1) is

$$W(\lambda) = Y_1 - Y_1^{-1}(P_1B - C^T)^T(\lambda I_n - A)^{-1}B \qquad (3.4.8)$$

where the symmetric matrix P_1 is any solution of

$$A^TP_1 + P_1A = -(P_1B - C^T)Y_0^{-1}(P_1B - C^T)^T. \qquad (3.4.9)$$

Proof. From (3.4.8)

$$\begin{aligned} W^T(-\lambda)W(\lambda) &= [Y_1 - B^T(-\lambda I - A^T)^{-1}(P_1B - C^T)Y_1^{-1}] \\ &\quad \times [Y_1 - Y_1^{-1}(P_1B - C^T)^T(\lambda I - A)^{-1}B] \\ &= Y_0 - (P_1B - C^T)^T(\lambda I - A)^{-1}B - B^T(-\lambda I - A^T)^{-1}(P_1B - C^T) \\ &\quad + B^T(-\lambda I - A^T)^{-1}(P_1B - C^T)Y_0^{-1}(P_1B - C^T)^T(\lambda I - A)^{-1}B \\ &= Y_0 - (P_1B - C^T)^T(\lambda I - A)^{-1}B - B^T(-\lambda I - A^T)^{-1}(P_1B - C^T) \\ &\quad - B^T(-\lambda I - A^T)^{-1}(A^TP_1 + P_1A)(\lambda I - A)^{-1}B \qquad \text{[using (3.4.9)]} \\ &= Y_0 + C(\lambda I - A)^{-1}B + B^T(-\lambda I - A^T)^{-1}C \\ &= Z(\lambda) + Z^T(-\lambda) \qquad \text{[using (3.4.7)]} \\ &= Y(\lambda) \qquad \text{as required.} \end{aligned}$$

Equation (3.4.9) is easily obtained from (3.4.6) with $\mathcal{J} = Y_0$. Notice that Theorem 3.27 guarantees the existence of at least one matrix P_1 satisfying (3.4.9).

For $W^{-1}(\lambda)$ from (3.4.8) to be analytic in $\mathrm{Re}(\lambda) > 0$ it is easily shown (see Exercise 3.4.8) that the characteristic roots of $A + BY_0^{-1}(B^TP_1 - C)$ must have negative real parts. It then turns out that the solution P_1 of (3.4.9) which satisfies this condition is unique. The equation (3.4.9) is a quadratic matrix equation of the type discussed in Section 5.3 (see Exercise 3.4.9 for a more convenient form) but care is needed because of the above requirement (see page 126). Anderson also describes modifications necessary if Y_0 is singular, and in a later paper [6] relates $W(\lambda)$ to the solution of a matrix Riccati differential equation (see page 117). For another application of Theorem 3.27 in control theory see [2].

We end this chapter by returning to Rosenbrock's formulation (Section 3). Using Theorem 3.27 he has proved:

THEOREM 3.29 [22]

(i) Let $P(\lambda)$ be a system matrix in first form corresponding to a real-rational transfer function matrix $G(\lambda)$.

(a) If $P(\lambda)$ is system similar to a positive-real matrix, then $G(\lambda)$ is positive-real.

(b) If $G(\lambda)$ is positive-real and $P(\lambda)$ has least order, then $P(\lambda)$ is system similar to a positive-real matrix.

(ii) If $P(\lambda)$ is in second form, then 'system similar' in (a) and (b) must be replaced by 'strictly system equivalent'.

A similar result [23] holds for *lossless positive-real* matrices, which are positive-real matrices satisfying the additional condition

$$Z(\lambda) + Z^T(-\lambda) = 0.$$

EXERCISES

3.4.1 [17] Let $Z_1(\lambda)$ and $Z_2(\lambda)$ be two positive-real $n \times n$ matrices. Show that $Z_1^T(\lambda)$, $Z_1(\lambda) + Z_2(\lambda)$ and $R^T Z_1(\lambda) R$, where R is an $n \times n$ real nonsingular matrix, are also positive-real.

3.4.2 Let $Y(\lambda) = Z(\lambda) + Z^T(-\lambda)$, where $Z(\lambda)$ is a positive-real rational matrix. Show that $Y(\lambda)$ is parahermitian, and that $Y(i\omega)$ is positive semidefinite Hermitian for all real ω, when $i\omega$ is not a pole of $Z(\lambda)$.

3.4.3 [30] Verify that if $W(\lambda)$ is a solution of (3.4.1), then so is $W_1(\lambda) = V(\lambda)W(\lambda)$, where $V(\lambda)$ is any $r \times r$ paraunitary matrix. Notice that $W_1(\lambda)$ will not in general have rank r.

3.4.4 Put $Y(\lambda) = A^T(-\lambda)A(\lambda)$ in Theorem 3.23, and hence show that any $n \times n$ real-rational matrix $A(\lambda)$ of rank r can be expressed as $V(\lambda)W(\lambda)$, where $V(\lambda)$ is paraunitary and $W(\lambda)$ is analytic for $\text{Re}\,(\lambda) > 0$.

3.4.5 [7] With

$$Y(\lambda) = \begin{bmatrix} \dfrac{-2\lambda^2+5}{(\lambda-1)(\lambda-2)(\lambda+1)(\lambda+2)} & \dfrac{-2\lambda^2-5\lambda+13}{(\lambda-1)(\lambda-2)(\lambda+3)(\lambda+5)} \\ \dfrac{-2\lambda^2+5\lambda+13}{(\lambda-3)(\lambda-5)(\lambda+1)(\lambda+2)} & \dfrac{-2\lambda^2+34}{(\lambda-3)(\lambda-5)(\lambda+3)(\lambda+5)} \end{bmatrix},$$

show that the steps of Davis's procedure give the following matrices:

(1) $T_1(\lambda) = \text{diag}\,[(\lambda + 1)(\lambda + 2), (\lambda + 3)(\lambda + 5)]$,

$$Y_1(\lambda) = \begin{bmatrix} -2\lambda^2 + 5 & -2\lambda^2 - 5\lambda + 13 \\ -2\lambda^2 + 5\lambda + 13 & -2\lambda^2 + 34 \end{bmatrix}.$$

(2) $$T_2(\lambda) = \begin{bmatrix} 1 & 0 \\ 0 & \dfrac{1}{\lambda + 1} \end{bmatrix} \quad T_3(\lambda) = \begin{bmatrix} 1 & 0 \\ \dfrac{-16}{3(\lambda + 1)} & 1 \end{bmatrix},$$

$$Y_2(\lambda) = \begin{bmatrix} -2\lambda^2 + 5 & \dfrac{26}{3}\lambda - \dfrac{41}{3} \\ \dfrac{-26}{3}\lambda - \dfrac{41}{3} & \dfrac{(26)^2}{18} \end{bmatrix}.$$

(3) $T_4(\lambda) = \begin{bmatrix} 1 & 3\lambda/13 \\ 0 & 1 \end{bmatrix}$, $Y_3 = \begin{bmatrix} 5 & -41/3 \\ -41/3 & 26^2/18 \end{bmatrix}$.

(4) $Y_4 = \begin{bmatrix} \sqrt{5} & 0 \\ -41/3\sqrt{5} & 1/\sqrt{5} \end{bmatrix}$.

Hence show that

$$W(\lambda) = \frac{1}{13\sqrt{5}} \begin{bmatrix} \dfrac{41\lambda + 65}{(\lambda+1)(\lambda+2)} & \dfrac{41\lambda + 169}{(\lambda+3)(\lambda+5)} \\ \dfrac{-3\lambda}{(\lambda+1)(\lambda+2)} & \dfrac{-3\lambda + 13}{(\lambda+3)(\lambda+5)} \end{bmatrix}.$$

3.4.6 [3] Prove that if A, B, and C satisfy (3.4.5) then $Z^T(\bar{\lambda}) + Z(\lambda)$ is positive semidefinite in Re $(\lambda) > 0$.

3.4.7 [5] Show that in Theorem 3.19, if $X(\lambda)$ is real-rational then

$$Z^*(i\omega) + Z(i\omega) = \tfrac{1}{2}[I_n + Z^*(i\omega)][I_n - X^*(i\omega)X(i\omega)][I_n + Z(i\omega)],$$

and hence deduce that condition (iii) of the theorem still holds if $Z(\lambda)$ is generalized positive-real, provided $i\omega$ is not a pole of $Z(\lambda)$.

3.4.8 Verify that the inverse of $W(\lambda)$ in equation (3.4.8) is

$$\{I_n + Y_0^{-1}(P_1B - C^T)^T[\lambda I_n - A - BY_0^{-1}(P_1B - C^T)^T]^{-1}B\}Y_1^{-1}.$$

3.4.9 Put $\alpha = A - BY_0^{-1}C$ in equation (3.4.9) and hence reduce this equation to

$$P_1BY_0^{-1}B^TP_1 + \alpha^TP_1 + P_1\alpha + C^TY_0^{-1}C = 0.$$

3.4.10 [17] A rational matrix $R(\lambda)$ is called *bounded-real* if in Re $(\lambda) > 0$, it is analytic; $\bar{R}(\lambda) = R(\bar{\lambda})$; and $I_n - R^*(\lambda)R(\lambda)$ is positive semidefinite. Show that the matrix $I_n - R(\lambda)$ is positive-real.

3.4.11 [22] If the first form matrix $P(\lambda)$ in equation (3.3.1) is positive-real, show that $\lambda I_n - A$ is also positive-real.

REFERENCES

1. ANDERSON, B. D. O., 'Development and applications of a system theory criterion for rational positive real matrices', *Proc. Fourth Allerton Conference on Circuit and System Theory*, University of Illinois, Urbana, 400–407 (1966).
2. ANDERSON, B. D. O., 'Stability of control systems with multiple nonlinearities,' *J. Franklin Inst.* **282,** 155–160 (1966).
3. ANDERSON, B. D. O., 'A system theory criterion for positive real matrices', *SIAM J. Control,* **5,** 171–182 (1967).
4. ANDERSON, B. D. O., 'An algebraic solution to the spectral factorization problem', *IEEE Trans. Aut. Control,* **AC-12,** 410–414 (1967).

5. ANDERSON, B. D. O. and MOORE, J. B., 'Algebraic structure of generalized positive real matrices', *SIAM J. Control*, **6**, 615–624 (1968).
6. ANDERSON, B. D. O., 'Quadratic minimization, positive real matrices and spectral factorization', *Tech. Rept. EE-6812*, University of Newcastle, New South Wales (1968).
7. DAVIS, M. C., 'Factoring the spectral matrix', *IEEE Trans Aut. Control*, **AC-8**, 296–305 (1963).
8. GANTMACHER, F. R., *Theory of matrices*, Vol. I, Chelsea Publishing Company, New York (1960).
9. GILBERT, E. G., 'Controllability and observability in multivariable control systems', *SIAM J. Control*, **1**, 128–150 (1963).
10. HEYMANN, M. and THORPE, J. A., 'Transfer equivalence of linear dynamical systems', *SIAM J. Control*, **8**, 19–40 (1970).
11. HO, B. L. and KALMAN, R. E., 'Effective construction of linear state-variable models from input/output functions', *Regelungstechnik*, **12**, 545–548 (1966).
12. KALMAN, R. E., 'Mathematical description of linear dynamical systems', *SIAM J. Control*, **1**, 152–192 (1963).
13. KALMAN, R. E., 'Irreducible realizations and the degree of a rational matrix', *SIAM J. Appl. Math.* **13**, 520–544 (1965).
14. KALMAN, R. E., 'On structural properties of linear, constant, multivariable systems', *IFAC Conference*, London (1966).
15. MAYNE, D. Q., 'Computational procedure for the minimal realisation of transfer-function matrices', *Proc. IEE*, **115**, 1363–1368 (1968).
16. MCMILLAN, B., 'Introduction to formal realizability theory—II', *Bell System Tech. J.* **31**, 541–600 (1952).
17. NEWCOMB, R. W., *Linear multiport synthesis*, McGraw-Hill, New York (1966).
18. PANDA, S. P. and CHEN, C. T., 'Irreducible Jordan form realization of a rational matrix', *IEEE Trans. Aut. Control*, **AC-14**, 66–69 (1969).
19. RIDDLE, A. C. and ANDERSON, B. D. O., 'Spectral factorization—computational aspects', *IEEE Trans. Aut. Control*, **AC-11**, 764–765 (1966).
20. ROSENBROCK, H. H., 'Efficient computation of least order for a given transfer function matrix', *Electronics Letters*, **3**, 413–414 (1967).
21. ROSENBROCK, H. H., 'The computation of minimal representations of a rational transfer-function matrix', *Proc. IEE*, **115**, 325–327 (1968).
22. ROSENBROCK, H. H., 'System matrices giving positive-real transfer function matrices', *Proc. IEE*, **115**, 328–329 (1968).

23. Rosenbrock, H. H., 'System matrices giving lossless positive-real transfer function matrices', *Proc. IEE*, **115**, 330–331 (1968).
24. Rosenbrock, H. H., 'McMillan forms from system matrices', *Electronics Letters*, **4**, 374–375 (1968).
25. Rosenbrock, H. H., 'Design of multivariable control systems using the inverse Nyquist array', *Proc. IEE*, **116**, 1929–1936 (1969).
26. Rosenbrock, H. H. and Rowe, A., 'Allocation of poles and zeros', *Proc. IEE*, **117**, 1879–1886 (1970).
27. Rosenbrock, H. H., *State-space and multivariable theory*, Nelson, London (1970).
28. Siljak, D., 'New algebraic criteria for positive realness', *Fourth Princeton Conference on Information Sciences and Systems*, Princeton University, New Jersey (1970).
29. Siljak, D., 'A criterion for nonnegativity of polynomial matrices with application to system theory', *Report No. NGR 05-017-010-039*, University of Santa Clara, California (1970).
30. Youla, D. C., 'On the factorization of rational matrices', *IRE Trans. Information Theory*, **IT-7**, 172–189 (1961).

4

Stability and Inertia

4.1 Real stability matrices and Liapunov theory

A *stability matrix* A is defined to be a matrix (with real or complex elements) of which all characteristic roots have negative real parts. In this section we shall take A to be real, leaving the complex case until Section 2. The importance of this definition is because of the well-known result:

THEOREM 4.1 The system of n linear differential equations with constant coefficients

$$\dot{x} = Ax \tag{4.1.1}$$

is asymptotically stable if and only if A is a stability matrix (x is an n-vector).

When A has a characteristic root with positive real part, (4.1.1) is unstable; if any root has zero real part then for stability the corresponding factors of the minimum polynomial of A must be linear. Readers unfamiliar with the precise definitions of stability and instability, or wishing for further details, should consult a standard textbook such as [24] or [38] (see also Appendix 3).

The problem of determining whether the characteristic roots of a given system matrix A satisfy the criterion of Theorem 4.1 has, of course, long been of importance in control engineering. The classical approach based on the characteristic polynomial of A has already been referred to in Chapter 2, but from a practical point of view the Routh, Hurwitz, and other methods all suffer from the difficulty of computing accurately the coefficients in the characteristic polynomial [82]. It is generally best, therefore (assuming access is available to a digital computer), to find the characteristic roots of A directly, using an efficient algorithm [82]. An added bonus of tackling the problem in this way is that the characteristic vectors of A can be found with little extra effort, so that the actual solution of (4.1.1) may be obtained, and the transient behaviour of the system determined.

However, from another point of view both the direct and classical approaches are unsatisfactory since they give no insight into the structure of stability matrices. This disadvantage is avoided when Liapunov's 'direct' method of stability analysis (see Appendix 3) is applied to linear systems, resulting in the following theorem which is fundamental for the remainder of this section.

THEOREM 4.2 The matrix A is a stability matrix if and only if, for any real positive definite symmetric matrix Q, the solution for the real symmetric matrix P of the *Liapunov matrix equation*

$$A^T P + PA = -Q \qquad (4.1.2)$$

is positive definite.

Several important points arising from the theorem should be noted:

(i) If P and Q satisfy the conditions of Theorem 4.2, the quadratic form $V = x^T Px$ is a 'Liapunov function' for (4.1.1) having derivative $-x^T Qx$.

(ii) If the solution for P of (4.1.2) is negative definite or indefinite then (4.1.1) is unstable.

(iii) It is of little value to choose P positive definite and then obtain the corresponding Q from (4.1.2), since unless Q turns out to be definite or semidefinite (which is unlikely), nothing can be inferred about the stability behaviour of (4.1.1) except that the quadratic form $x^T Px$ is not a Liapunov function.

(iv) If Q in (4.1.2) is chosen positive semidefinite, then if P turns out to be positive definite Liapunov's theory establishes that (4.1.1) is certainly stable. However, in view of incorrect statements which have appeared in print [62, 63, 80] we must stress that positive definiteness of P is neither a necessary nor a sufficient condition for *asymptotic* stability. The following simple examples of matrices satisfying equation (4.1.2) illustrate this point.

(a) $A = \begin{bmatrix} 1 & -3 \\ 2 & -4 \end{bmatrix}$ has characteristic roots $-1, -2$,

$P = \begin{bmatrix} 1 & -1 \\ -1 & 1 \end{bmatrix}$, $Q = \begin{bmatrix} 2 & -2 \\ -2 & 2 \end{bmatrix}$ are both positive semidefinite.

(b) $A = \begin{bmatrix} 0 & 1 \\ 0 & -1 \end{bmatrix}$ has characteristic roots $0, -1$,

$P = \begin{bmatrix} 1 & 1 \\ 1 & 2 \end{bmatrix}$ is positive definite,

$Q = \begin{bmatrix} 0 & 0 \\ 0 & 2 \end{bmatrix}$ is positive semidefinite.

However, when Q is positive semidefinite it can always be written in the form RR^T, where R is an $n \times m$ matrix with rank equal to that of Q. An important theorem given in slightly different forms by two authors [1, 39] may then be stated as follows:

THEOREM 4.3 If A is a stability matrix then the solution for the real symmetric matrix P of the equation

$$A^TP + PA = -RR^T$$

is unique and positive semidefinite. Further, if the pair $[A, R^T]$ is observable†, i.e., if $\operatorname{rank}[R, A^TR, (A^T)^2R, \ldots, (A^T)^{n-1}R] = n$, or equivalently if $R^Te^{A^t}x = 0$ for all t implies $x = 0$, then P is positive definite.

For further remarks on the semidefinite case see [78] and Section 2 of this chapter.

(v) Since P is symmetric, (4.1.2) represents $\frac{1}{2}n(n + 1)$ linear equations for the $\frac{1}{2}n(n + 1)$ unknown elements of P. By Theorem 1.9, the solution of (4.1.2) will be unique provided A^T and $-A$ have no common characteristic root (i.e. provided there are no roots λ_i, λ_j, of A such that $\lambda_i + \lambda_j = 0$). Much effort and ingenuity has been devoted to finding efficient methods of solving (4.1.2) and we shall make some comments on these later. When A takes certain special forms the solution of (4.1.2) can be written down at once (see Exercises 4.1.1, 4.1.2); however, since in general the computational effort increases in proportion to at least n^3 it must be admitted that the usefulness of Theorem 4.2 as a practical tool in the stability analysis problem is limited.

(vi) Some properties of the transformation $P \to A^TP + PA$ represented by (4.1.2) (including the complex case) have been investigated by Givens [25]. The slightly more general expression $A^TP^T + PA$ where P is no longer symmetric has been studied by Taussky and Wielandt [73]. Trampus [77] deals with the yet more general transformation $P \to AP + PB$.

(vii) In 1962 Parks [53] proved the Routh–Hurwitz conditions directly using Liapunov's method, thus avoiding the complex integral calculus needed in the usual proofs. Since then several other links between the classical and Liapunov theorems have been investigated, and Anderson [2] gives a good account of the interrelationships (see also [54]).

---

† *See Appendix* 2.

(viii) For an important generalization of the Liapunov matrix equation to the much-studied Luré problem in control theory, see [33].

As has already been indicated, the novelty of using Liapunov's approach as compared with those of Hermite, Routh, Hurwitz, and others lies in the fact that we can deal directly with the system matrix itself, instead of the coefficients in its characteristic polynomial. This means that the whole apparatus of matrix techniques is at our disposal; accordingly, we shall see that the importance of Theorem 4.2 lies in the way in which it can be applied and generalized within the realm of matrix theory (see [76] for mention of abstract applications).

We begin our development by noticing from the proof of Theorem 1.9 (see also Appendix 1) that equation (4.1.2) can be written

$$(A^T \otimes I_n + I_n \otimes A^T)p = -q, \tag{4.1.3}$$

where p and q are column n^2-vectors formed from the rows of P and Q respectively taken in order. The matrix on the left in (4.1.3) is of order n^2 and has characteristic roots $\lambda_i + \lambda_j$, $i, j = 1, 2, \ldots, n$. As might be expected from the symmetry of P, redundant equations and variables can be removed from (4.1.3) to give

$$B\rho = -\phi, \tag{4.1.4}$$

where B is of order $\frac{1}{2}n(n+1)$ and ρ and ϕ are column vectors composed of the elements on and above the principal diagonals of P and Q respectively—for example, $\rho = [p_{11}, p_{12}, p_{22}, p_{13}, p_{23}, p_{33}, \ldots]^T$. The matrix B is easily constructed from A, and a straightforward method of doing this has been given by MacFarlane [40], who also shows in the same paper how performance integrals associated with (4.1.1) can be obtained by solving (4.1.2) (see Exercises 4.1.13–4.1.17). Fuller [23] has also studied the relationship between equations (4.1.3) and (4.1.4).

It turns out, however, that the order of the matrix which has to be inverted in order to solve (4.1.2) can be reduced by a further amount n. We proceed as follows [4]: write (4.1.2) in the form

$$(PA + \tfrac{1}{2}Q) + (PA + \tfrac{1}{2}Q)^T = 0,$$

so

$$PA + \tfrac{1}{2}Q = S, \tag{4.1.5}$$

where S is a real skew symmetric matrix. Hence

$$P = (S - \tfrac{1}{2}Q)A^{-1}$$

provided A is nonsingular, and S is determined from the condition $P^T = P$, which reduces to

$$A^TS + SA = \tfrac{1}{2}(A^TQ - QA). \tag{4.1.6}$$

Since the principal diagonal of S is identically zero, (4.1.6) represents $\frac{1}{2}n(n-1)$ equations for the $\frac{1}{2}n(n-1)$ unknown elements of S, which can be written in a similar fashion to (4.1.4) as

$$\beta s = -\gamma.$$

The matrix β (of order $\frac{1}{2}n(n-1)$) is easily found from B (see [12]) and $s = [s_{12}, s_{13}, s_{23}, s_{14}, s_{24}, \ldots]^T$. Fuller, in the paper [23] mentioned previously, has also obtained a matrix corresponding to β by a different route, using the 'bialternate product' of A with itself.

Once the skew symmetric matrix has been introduced a second application becomes at once apparent from (4.1.5). We have (replacing P^{-1} by P_0, Q by $2Q_0$)

THEOREM 4.4 [5] The matrix

$$A = P_0(S_0 - Q_0) \tag{4.1.7}$$

is a stability matrix for any real symmetric positive definite matrices P_0 and Q_0 and real skew symmetric matrix S_0. Further, the real parts of the characteristic roots of A lie between the greatest and least roots of the matrix $-P_0Q_0$.

The bounds in the second part of the theorem are obtained using a result due to Vogt [79] (see Exercise 4.1.5); notice that these bounds are independent of S_0. MacFarlane [42] has made an interesting application of the theorem to a problem in electrical network theory.

In this general expression for a stability matrix which we now have it would be very useful if we could choose P_0, Q_0 and S_0 so as to determine the actual magnitudes of some or all of the real parts of the roots of A in (4.1.7). This would mean that we could construct a linear system (4.1.1) with transient response more precisely determined than by the bounds of the theorem. Unfortunately no progress has at present been made in this direction but it would seem a worthwhile area for research.

We can, however, by a similar argument derive very easily the following theorem on changes in the elements of a given stability matrix which do not make it unstable:

THEOREM 4.5 [5] If A is a stability matrix then so is $A + B$, where

$$B = (S_1 - Q_1)P, \tag{4.1.8}$$

P being the solution of (4.1.2) and S_1 and Q_1 being arbitrary skew and positive semidefinite matrices respectively.

This result follows at once on substitution of $A + B$ into (4.1.2), and shows that for a given matrix A solution of the Liapunov matrix equation can be used to provide further potentially valuable information about the associated system of differential equations. Notice that if Q_1 is definite then B is itself a stability matrix by virtue of Theorem

4.4. Of course (4.1.8) represents only a sufficient and not a necessary condition on B for $A + B$ to be a stability matrix, but nevertheless the ease with which the expression can be written down illustrates the power of applying Liapunov's method to obtain results in matrix theory without complex proofs. Indeed, it is hard to see how either (4.1.7) or (4.1.8) could be obtained by conventional methods; Theorem 4.5 is especially of interest since in general it is difficult to say much about changes in the characteristic roots of a matrix if the elements alter by finite amounts [75].

The basis of the argument which has been used to derive Theorems 4.4 and 4.5 is simply to find linear systems having the same quadratic form Liapunov function. This idea can be extended to deal with non-linear systems of differential equations and to some types of control systems. Further discussion is outside the scope of this book, and we refer the interested reader to [8], [10] or [12].

If instead of linear differential equations we consider discrete-time systems

$$x(t_{k+1}) = \alpha x(t_k), \qquad (k = 0, 1, 2, \ldots) \tag{4.1.9}$$

where α is an $n \times n$ real constant matrix, the condition for asymptotic stability becomes that $\alpha^n \to 0$ as $n \to \infty$ i.e. that α be a *convergent* matrix. This implies

THEOREM 4.6 The system (4.1.9) is asymptotically stable if and only if all the characteristic roots of α have modulus less than unity.

The result corresponding to Theorem 4.2 derived from Liapunov theory is

THEOREM 4.7 The system (4.1.9) is asymptotically stable if and only if for any positive definite symmetric matrix M the solution for the real symmetric matrix R of

$$\alpha^T R \alpha - R = -M \tag{4.1.10}$$

is positive definite.

In fact there is a direct relationship between the equations (4.1.2) and (4.1.10) which is easily made apparent by putting

$$A = (\alpha - I_n)(\alpha + I_n)^{-1} \tag{4.1.11}$$

into (4.1.2), giving

$$\alpha^T P \alpha - P = -\tfrac{1}{2}(\alpha^T + I_n)Q(\alpha + I_n). \tag{4.1.12}$$

We are now in a position to indicate briefly some methods which have been suggested for solving equation (4.1.2) (or equivalently equation (4.1.10)). These fall broadly into three classes:

(i) Solving directly the linear equations represented by (4.1.2) for the unknown elements of P, as expressed for example in equation (4.1.4). The main disadvantage is that the number of equations and unknowns is proportional to n^2.

(ii) Methods which rely on converting A by means of a similarity transformation into a suitable standard form which enables the resulting equations to be solved more easily. Particularly useful are the companion form [49] (see Appendix 1) and Schwarz form [32] (see Exercise 4.1.1). Disadvantages, which are quite severe, are that A is required to be nonderogatory, and that rounding errors in the transformation can be relatively large.

(iii) The most successful approaches seem to be those which express the solution P in terms of either a finite matrix series [16] or a convergent infinite matrix series. In the latter method the transformation (4.1.11) plays a fundamental role, since it is easy to write down the solution of equation (4.1.12) (see Exercise 4.1.9), and the difficulties encountered in (i) and (ii) do not arise.

Details of solution methods, many of which are also applicable to more general linear matrix equations of the form (1.2.4), are outside the scope of this book and we refer the reader to [12] for a full treatment.

Further study of the transformation (4.1.11) and the relationship between equations (4.1.2) and (4.1.12) has been carried out by Taussky [74], Stein [68–70] and Power [55]. Other results and references on stability matrices and related topics can be found in [3], [58], [59], [64–67], [71, 75] and the review [76].

We close this section by considering again the linear control system of Section 1.4 described by

$$\dot{x} = Ax + Bu, \tag{4.1.13}$$

where A is $n \times n$, B is $n \times l$ and u is the vector of controlled variables. For various reasons, the 'free' or uncontrolled system $u = 0$ may well be unstable and the aim is then to choose u so that the resulting system is asymptotically stable. Of fundamental importance in control engineering is the idea of 'linear feedback':

$$u = Kx,$$

where K is an $l \times n$ constant matrix (the 'gain' matrix). When applied to (4.1.13) this gives

$$\dot{x} = (A + BK)x,$$

the so-called 'closed-loop' system ((4.1.13) being referred to as the 'open-loop' situation). The significance of this principle is because of the following result which provides the foundation for linear theory:

THEOREM 4.8 [83] Let Λ_n be an arbitrary set of n complex numbers such that any which do not have zero imaginary part occur in conjugate pairs. Then, given real matrices A and B, there exists a real matrix K such that the characteristic roots of $A + BK$ are the set Λ_n if and only if the pair $[A, B]$ is controllable; i.e. if and only if

$$\text{rank}\,[B, AB, A^2B, \ldots, A^{n-1}B] = n.$$

Wonham's original proof [83] of the general case is rather complicated, but a simpler proof has since been produced [29]. The case when B is a column vector ($l = 1$) is much more easily dealt with (see Exercises 4.1.11 and 4.1.12).

Theorem 4.8 thus tells us that simply by choosing each control variable u_i to be a suitable linear combination of the state variables $x_1, x_2, \ldots, x_n$, the characteristic roots of the resulting closed loop system can be predetermined; thus, for example, they can be made to have negative real parts. Of course the theorem only guarantees the existence of a suitable matrix K and the problem of how to choose K (and possibly B also) in some optimal fashion, although much investigated, is still not satisfactorily solved [60]. The theorem has recently been extended [22] to the case of linear output feedback $u = Ky$, where $y = Cx$ is an m-vector of output variables: provided rank $C = m \leqslant n$, A is nonderogatory and $[A, B]$ is controllable, a matrix K exists such that m characteristic roots of the closed-loop matrix $A + BKC$ are arbitrarily close to the members of a set Λ_m.

EXERCISES

4.1.1 The matrix

$$W = \begin{bmatrix} 0 & 1 & 0 & \cdots & \cdot & \cdot \\ -w_1 & 0 & 1 & \cdots & \cdot & \cdot \\ 0 & -w_2 & 0 & \cdots & \cdot & \cdot \\ \cdot & \cdot & \cdot & \cdots & 0 & 1 \\ \cdot & \cdot & \cdot & \cdots & -w_{n-1} & -w_n \end{bmatrix}$$

is said to be in *Schwarz form*. Show that (4.1.2) with $A = W$ and $Q = \text{diag}\,[0, 0, \ldots, 0, 1]$ has a solution with P diagonal, and hence obtain a necessary and sufficient condition for W to be a stability matrix (it can be shown that semidefinite Q causes no loss of generality in this case).

Because of the simplicity of P in this case much effort has been devoted to putting a general matrix into Schwarz form. For more details, see [12].

4.1.2 [9] Let T be an $n \times n$ tridiagonal matrix, so $t_{ij} = 0$ if $|i - j| \geqslant 2$. Take P and Q in (4.1.2) to be diagonal and hence show that sufficient conditions for T to be a stability matrix are $t_{ii} < 0$, $i = 1, 2, \ldots, n$ and $t_{i,\,i+1}\, t_{i+1,\,i} < 0$, $i = 1, 2, \ldots, n - 1$.

4.1.3 [8] Solve equation (4.1.2) with

$$A = \begin{bmatrix} 4 & 6 \\ -5 & -7 \end{bmatrix}, \quad Q = \begin{bmatrix} \lambda_1 & 0 \\ 0 & \lambda_2 \end{bmatrix}$$

and hence verify that A is a stability matrix.

Take $S_1 = 12 \begin{bmatrix} 0 & \alpha \\ -\alpha & 0 \end{bmatrix}$, $Q_1 = 12 \begin{bmatrix} q_1 & 0 \\ 0 & q_2 \end{bmatrix}$ in (4.1.8) and obtain the expression for B. Notice the wide freedom of choice in the parameters which keep $A + B$ a stability matrix.

4.1.4 Show that in equation (4.1.7), trace $(P_0 S_0) = 0$. Hence deduce that $\Sigma\lambda_i(A) = -\Sigma\lambda_i(P_0 Q_0)$.

Similarly in equation (4.1.8) show that $\operatorname{tr} B = -\Sigma\lambda_i(Q_1 P)$, and hence deduce that the principal diagonal of B cannot be identically zero except when Q_1 is identically zero.

4.1.5 [79] If P and Q satisfy (4.1.2) and x_i is a characteristic vector of A corresponding to a characteristic root λ_i, show that

$$\operatorname{Re}(\lambda_i) \equiv \alpha_i = -x_i^* Q x_i / 2x_i^* P x_i.$$

Hence, using Rayleigh's theorem [15, p. 112], show that if μ_1 and μ_2 are the smallest and largest roots of $\frac{1}{2}QP^{-1}$ and α_1, α_n are the largest and smallest α_i, then $\mu_1 \leqslant -\alpha_1$ and $\mu_2 \geqslant -\alpha_n$.

4.1.6 Using the previous exercise, obtain the bounds in Theorem 4.4.

4.1.7 [1] Use Theorem 1.9 to show that the only matrices which commute with $\begin{bmatrix} A & 0 \\ 0 & -A^T \end{bmatrix}$, where A is a stability matrix, are of the form $\begin{bmatrix} T_1 & 0 \\ 0 & T_2 \end{bmatrix}$, where T_1 and T_2^T commute with A.

4.1.8 [32] Show that the real parts of the characteristic roots of A are less than some real scalar k if and only if the solution of

$$A^T P + PA - 2kP = -Q$$

is positive definite for any symmetric positive definite Q.

4.1.9 Verify that the matrix equation

$$P - \alpha^T P \alpha = M$$

(see (4.1.12)) has solution

$$P = M + \alpha^T M \alpha + (\alpha^T)^2 M \alpha^2 + (\alpha^T)^3 M \alpha^3 + \cdots$$

Assume that α is similar to a diagonal matrix and hence show that the series for P converges provided α is a convergent matrix. (See [57] for a more general convergence proof.)

4.1.10 [7] Consider the linear system (4.1.1) where the elements of A depend upon some scalar parameter θ. Let $y = \partial x/\partial\theta$, $A_\theta = \partial A/\partial\theta$ and show that

$$\begin{bmatrix} \dot{x} \\ \dot{y} \end{bmatrix} = \begin{bmatrix} A & 0 \\ A_\theta & A \end{bmatrix} \begin{bmatrix} x \\ y \end{bmatrix}.$$

Hence deduce that if (4.1.1) is asymptotically stable for some value θ_0 of θ then $(\partial x/\partial\theta)_{\theta_0} \to 0$ as $t \to \infty$. By a similar argument show that $(\partial^2 x/\partial\theta^2)_{\theta_0}, (\partial^3 x/\partial\theta^3)_{\theta_0}, \ldots$, all $\to 0$ as $t \to \infty$.

4.1.11 Consider the system $\dot{x} = Ax + bu$, where A is $n \times n$ and b is $n \times 1$. Show that there exists a transformation $y = Tx$, where

$$T = \begin{bmatrix} t \\ tA \\ \cdot \\ \cdot \\ \cdot \\ tA^{n-1} \end{bmatrix}$$

which puts A into companion form and b into $[0, 0, \ldots, 0, 1]^T$ if and only if rank $[b, Ab, \ldots, A^{n-1}b] = n$.

4.1.12 Using the result of the previous exercise, complete the proof of Theorem 4.8 in this case.

4.1.13 Show that $d(x^T Px)/dt = x^T Qx$ where $x(t)$ is the solution of (4.1.1) and $A^T P + PA = Q$. Hence show that

$$\int_0^t x^T Qx \, dt = x^T(t)Px(t) - x^T(0)Px(0),$$

and deduce that if A is a stability matrix

$$\int_0^\infty x^T Qx \, dt = -x^T(0)Px(0).$$

4.1.14 If A is given by equation (4.1.7), using the result of the previous exercise show that

$$\int_0^\infty x^T Q_0 x \, dt = \tfrac{1}{2} x^T(0) P_0^{-1} x(0).$$

Notice that this is independent of S_0.

4.1.15 [43] If $x(t)$ is the solution of (4.1.1) show that $d(x \otimes x)/dt = B(x \otimes x)$, where $B = A \otimes I + I \otimes A$. Hence show that if A is a stability matrix,

$$\int_0^\infty (x \otimes x) \, dt = -B^{-1} x(0) \otimes x(0).$$

4.1.16 [40] By differentiating $tx^T Px$ with respect to t and using the result of Exercise 4.1.13, show that when A is a stability matrix,

$$\int_0^\infty tx^T Qx \, dt = (-1)^2 x^T(0) P_1 x(0),$$

where $A^T P_1 + P_1 A = P$. By repeating this argument show that

$$\int_0^\infty t^r x^T Qx \, dt = (-1)^{r+1} r! \, x^T(0) P_r x(0)$$

where $A^T P_s + P_s A = P_{s-1}$, $s = 0, 1, 2, \ldots, r \qquad (P_0 \equiv P)$.

4.1.17 [43] As in the previous exercise, differentiate $tx \otimes x$ and use Exercise 4.1.15 to show that

$$\int_0^\infty tx \otimes x\,dt = B^{-2}x(0) \otimes x(0),$$

$$\int_0^\infty t^r x \otimes x\,dt = (-1)^{r+1} r!\, B^{-(r+1)}x(0) \otimes x(0).$$

See MacFarlane's papers [41] and [12] for details of how more general system functionals can be evaluated, as in the five preceding exercises, without explicit determination of $x(t)$. See also [6] and [11] for further algebraic detail.

4.1.18 Show that if $A^T Q = QA$ with Q positive definite then all the characteristic roots of A must be real [15, p. 67]. Notice that S in equation (4.1.6) then reduces to a zero matrix. (See also [20].)

4.1.19 [76] Verify that the characteristic roots of

$$A(t) = \begin{bmatrix} -4 & 3e^{-8t} \\ -e^{8t} & 0 \end{bmatrix}$$

are both real and negative, but that the solution of $\dot{x} = A(t)x$ diverges as $t \to \infty$.

This illustrates the difficulty of dealing with matrices having non-constant elements.

4.1.20 Using Theorem 4.7 show that the matrix $(T + R)(T - R)^{-1}$ is convergent for any positive definite symmetric R and any T having negative definite symmetric part.

4.2 Inertia of complex matrices

In most cases when the system of differential equations (4.1.1) arises from a physical situation the elements of the matrix A are real. However, this is not always so and in this section we allow A to be complex. First, define the *inertia* of A to be the ordered integer triple $(\pi(A), \nu(A), \delta(A)) = \text{In}\,(A)$, where π, ν, δ are the numbers of characteristic roots of A having positive, negative and zero real parts respectively. Thus a stability matrix (real or complex) has inertia $(0, n, 0)$. In these terms Sylvester's well-known inertia theorem becomes:

THEOREM 4.9 If H is Hermitian and P is nonsingular,

$$\text{In}\,(H) = \text{In}\,(P^*HP).$$

This result has been sharpened by Ostrowski [51] as follows:

THEOREM 4.10 If the characteristic roots of H are

$$\lambda_1 \leqslant \lambda_2 \leqslant \cdots \leqslant \lambda_n$$

and those of P^*HP are $\mu_1 \leqslant \mu_2 \leqslant \cdots \leqslant \mu_n$, then $\mu_i = \theta_i \lambda_i$, $i = 1, 2, \ldots, n$, where the θ_i lie between the smallest and largest characteristic roots of the positive definite matrix P^*P.

Ostrowski and Schneider [52] and Taussky [72] have proved the following generalization of Theorem 4.2:

THEOREM 4.11 For a given complex matrix A there exists a unique Hermitian H satisfying the equation

$$A^*H + HA = 2C, \qquad (4.2.1)$$

where C is positive definite Hermitian, if and only if the characteristic roots λ_i of A are such that $\bar{\lambda}_i + \lambda_j \neq 0$, all i, j, and then $\text{In}(A) = \text{In}(H)$.

We now make some relevant comments:

(i) The first part of the theorem is a consequence of Theorem 1.9. Ostrowski and Schneider [52] also show that there exists a Hermitian H such that $A^*H + HA$ is positive definite if and only if $\delta(A) = 0$ (and $\text{In}(H)$ is still equal to $\text{In}(A)$); and that if $\delta(A) = 0$ then it is always possible to find a Hermitian H such that $\text{In}(A^*H + HA)$ is equal to any given triple $(\pi, \nu, 0)$.

(ii) When C is semidefinite Lehnigk [39] has shown that Theorem 4.3 still holds with transpose replaced by conjugate transpose. Wigner and Yanase [81] give some theorems when A is also Hermitian but C not necessarily positive definite.

(iii) Liapunov's Theorem 4.2 is seen to be a special case of Theorem 4.11 (with a trivial change of sign). With this latter result we therefore have an extension of Liapunov's linear theory in a similar way to that whereby the classical Routh, Hurwitz, and other criteria can be modified to determine the number of roots of a polynomial in *each* half-plane (see Chapter 2, Section 3). Lehnigk [39] in fact derives Hermite's theorem (Theorem 2.12) directly, remarking, 'with such Liapunov-type proofs available, it is possible to treat every aspect of the stability theory of linear motions by means of the tools provided by Liapunov's direct method'.

(iv) It follows easily as in Theorem 4.4 that if

$$A = H^{-1}(S + C), \qquad (4.2.2)$$

then $\text{In}(A) = \text{In}(H)$ for any skew Hermitian S and Hermitian H [31]. Thus we can construct a matrix with predetermined inertia; in fact it is also possible to find bounds on the real parts α_i of the characteristic roots of A. It is convenient to begin by ordering the characteristic roots of the matrix $C^{-1}H$ according to

$$\min_i \mu_i(C^{-1}H)$$
$$= \mu_1 \leqslant \mu_2 \leqslant \cdots \leqslant \mu_{n_1} < 0 < \mu_{n_1+1} \leqslant \cdots \leqslant \mu_n = \max_i \mu_i(C^{-1}H).$$

Notice that $\operatorname{In}(C^{-1}H) = \operatorname{In}(H)$ since C is positive definite, so $\operatorname{In}(A) = \operatorname{In}(C^{-1}H) = \operatorname{In}(H^{-1}C)$. Then, writing $\gamma_i = 1/\mu_i$, the following result is easily proved:

THEOREM 4.12 [31] For the matrices in (4.2.2),

$$\mu_1 \leqslant 1/\alpha_i \leqslant \mu_n.$$

Further, if

(a) $\operatorname{In}(A) = (n, 0, 0)$
then $0 < \gamma_n \leqslant \gamma_{n-1} \leqslant \cdots \leqslant \gamma_1$
and $\gamma_n \leqslant \alpha_i \leqslant \gamma_1$;

(b) $\operatorname{In}(A) = (0, n, 0)$
then $\gamma_n \leqslant \gamma_{n-1} \leqslant \cdots \leqslant \gamma_1 < 0$
and again $\gamma_n \leqslant \alpha_i \leqslant \gamma_1$;

(c) $\operatorname{In}(A) = (n - n_1, n_1, 0)$
then $\gamma_{n_1} \leqslant \gamma_{n_1-1} \leqslant \cdots \leqslant \gamma_1 < 0 < \gamma_n \leqslant \gamma_{n-1} \leqslant \cdots \leqslant \gamma_{n_1+1}$
and $\alpha_i \geqslant \gamma_n, \quad i \in \{j \mid \alpha_j > 0\}$
$\alpha_i \leqslant \gamma_1, \quad i \in \{j \mid \alpha_j < 0\}$.

It is also easy to obtain the result corresponding to Theorem 4.5. For further discussion and some examples see [31]. It should be noted, however, that as in the special case of stability matrices (Theorem 4.4) it is not yet known if it is possible to choose the various matrices in (4.2.2) so as to determine completely the real parts of some of the characteristic roots of A.

(v) Theorem 4.7 also generalizes to the complex case [68], transpose in (4.1.10) being replaced by conjugate transpose.

When C in equation (4.2.1) is only positive semidefinite the relationship between $\operatorname{In}(H)$ and $\operatorname{In}(A)$ is much more complicated. An important theorem is:

THEOREM 4.13 [18] For any A there exists a nonsingular H which makes C in (4.2.1) positive semidefinite. Further, there exists a positive definite H if and only if (a) $\nu(A) = 0$ and (b) all elementary divisors of pure imaginary characteristic roots (if any) of A are linear.

Carlson and Schneider [18] also give the best possible bounds for $\pi(H)$, $\nu(H)$ and rank C when condition (b) of Theorem 4.13 holds. Other special cases are considered in the same paper, and Carlson [19] discusses the case when A has roots α and $-\bar{\alpha}$. He also uses similar methods to derive results concerning complex matrices having real characteristic roots [20]. Other inertia theorems may be found in a paper by one of Carlson's associates [30] and the excellent review by Taussky [76] should also be consulted.

Haynsworth [27, 28] has obtained several results on the inertia of partitioned Hermitian matrices. We give one here which seems especially interesting because of its connection with Schur's well-known expression for the determinant of a partitioned matrix. First we remark that if A (of rank at least k) is Hermitian it is easy to show that there exists a permutation matrix† R such that $R^T AR$ has a nonsingular matrix of order k in the upper left corner. Thus we can assume without loss of generality that

$$A = \begin{bmatrix} A_{11} & A_{12} \\ A_{12}^* & A_{22} \end{bmatrix}$$

where A_{11} is nonsingular Hermitian of order k. Then Schur's formula [24, Vol. I, p. 46] gives

$$\det A = \det A_{11} \det B,$$

where $B = A_{22} - A_{12}^* A_{11}^{-1} A_{12}$. Haynsworth calls B the 'Schur complement' of A_{11} and obtains the unexpectedly simple result:

THEOREM 4.14 $\text{In}(A) = \text{In}(A_{11}) + \text{In}(B)$.

Proof. Let

$$L = \begin{bmatrix} I_k & -A_{11}^{-1}A_{12} \\ 0 & I_{n-k} \end{bmatrix}.$$

It is straightforward to verify that

$$L^*AL = \begin{bmatrix} A_{11} & 0 \\ 0 & B \end{bmatrix}. \tag{4.2.3}$$

Since $\text{In}(L^*AL) = \text{In}(A)$ (Theorem 4.9) and since the characteristic roots of the matrix on the right in (4.2.3) are just those of A_{11} and B, the result follows.

Clearly partitioning can be of practical value in stability and inertia problems by reducing the order of matrices to be dealt with, but at present no results are known for non-Hermitian matrices.

EXERCISES

4.2.1 Show that if A in equation (4.2.1) is tridiagonal then H and C can be taken diagonal provided $a_{i,i+1} = k_i \bar{a}_{i+1,i}$, where k_i is real ($i = 1, 2, \ldots, n-1$).

† *See Appendix* 1.

This can be used to construct a tridiagonal matrix with given inertia [31].

4.2.2 If H is Hermitian and nonsingular, and $A + A^*$ is positive definite, use equation (4.2.1) to show that In $(HA) =$ In (H). (The result also holds when H is singular—see [52].)

4.2.3 [28] Let the Hermitian matrix H be

$$H = \begin{bmatrix} H_1 & H_2 \\ H_2^* & 0 \end{bmatrix},$$

where each of the submatrices is $n \times n$ and H_1 and H_2 are nonsingular. Show that In $(H) = (n, n, 0)$, using Theorem 4.14.

4.2.4 [81] If A is Hermitian, H positive definite Hermitian and $AH + HA = 0$, show that $A = 0$.

4.2.5 [81] If α and β are positive definite Hermitian and $\alpha^2 > \beta^2$ (i.e. $\alpha^2 - \beta^2$ is positive definite) show that $\alpha > \beta$. (Hint: take $2C = \alpha^2 - \beta^2$ in equation (4.2.1).)

4.2.6 [52] If H is Hermitian and $A^*H + HA$ is positive definite, show that H is nonsingular.

4.3 Qualitative stability

A situation in mathematical economics which has received considerable attention is when no quantitative information is available. More specifically, we shall consider here the stability problem for linear systems where only the signs (including zeros) of the elements of matrices are known [47, 56].

Two real $n \times n$ matrices $A = [a_{ij}]$ and $B = [b_{ij}]$ are said to be *sign similar* if sgn $(a_{ij}) =$ sgn (b_{ij}) or $a_{ij} = b_{ij} = 0$, all i, j. Let $Q(A)$ denote the set of matrices which are sign similar to a given matrix A. Then A is called *sign stable* (or *qualitatively stable*) if and only if every member of $Q(A)$ is a stability matrix. In other words, A is sign stable if it is a stability matrix whatever the value of its elements, provided the sign pattern is preserved. If at least one member of $Q(A)$ is a stability matrix, A is said to be *potentially stable*. Clearly, multiplying any matrix by a positive diagonal matrix $D =$ diag $[d_1, d_2, \ldots, d_n]$, all $d_i > 0$, does not alter its sign pattern. For this reason, defining a matrix A to be *D-stable* if DA is a stability matrix for any positive diagonal matrix D, it then follows that any sign stable matrix is also D-stable. A sufficient condition for D-stability is easily derived:

THEOREM 4.15 If there exists a positive diagonal matrix B such that $A^TB + BA$ is negative definite, then A is D-stable.

Proof. If $A^TB + BA = -Q$, then

$$(AD)^TBD + DB(AD) = -DQD,$$

for any positive diagonal D. Since BD and DQD are both positive definite, AD is a stability matrix by Theorem 4.2.

Using the Routh–Hurwitz conditions, Quirk and Ruppert [56] have obtained a necessary condition for D-stability:

THEOREM 4.16 If A is D-stable, then no principal minor of A of odd order is positive, no principal minor of even order is negative and at least one non-zero principal minor of every order exists.

At present no necessary *and* sufficient condition for D-stability is known.

Closely related to D-stable matrices are *totally stable* matrices, which are defined by the condition that every principal submatrix be D-stable. Metzler [48] has obtained

THEOREM 4.17 A necessary condition for A to be totally stable is that every principal minor of A of even order is positive and every principal minor of odd order is negative (these conditions on minors are referred to by economists as the *Hicks conditions*—notice that the conditions of Theorem 4.16 are 'almost' Hicksian).

Clearly A totally stable $\Rightarrow A$ D-stable $\Rightarrow A$ stable; further, if all the diagonal elements of A are negative, then A sign stable $\Rightarrow A$ totally stable [56]. Other definitions may also be mentioned: let R be real symmetric and H Hermitian—then A is called R- (or H-) *stable* if AR (or AH) is stable whenever R (or H) is positive definite. Similarly A is *H-semistable* if $\nu(AH) = 0$ whenever H is positive definite [52]. Carlson has derived a necessary and sufficient condition for H-stability:

THEOREM 4.18 [21] Let

$$R(A) = (A + A^*)/2, \quad Im\,(A) = (A - A^*)/2i.$$

Then A is H-stable if and only if

(i) $R(A)$ is positive semidefinite
(ii) $x^*R(A)x = 0 \Rightarrow x^*Im\,(A)x = 0$
(iii) A is nonsingular.

If A is real, the conditions are then for R-stability and (ii) is replaced by

(ii′) for each pair of real vectors ξ, η for which

$$R(A)\xi = 0 = R(A)\eta, \qquad \text{we have } \xi' A\eta = 0.$$

Quirk and Ruppert [56] prove two theorems on sign stability, the first for matrices with all elements on the principal diagonal negative, and the second a generalization:

THEOREM 4.19 Let A be an $n \times n$ real indecomposable† matrix. Then A is sign stable if and only if

(i) $a_{ij}a_{ji} \leqslant 0, \quad i \neq j$
(ii) $i_1 \neq i_2 \neq \cdots \neq i_m$,
$a_{i_1i_2} \neq 0, a_{i_2i_3} \neq 0, \ldots, a_{i_{m-1},i_m} \neq 0 \Rightarrow a_{i_mi_1} = 0$,
for any $m > 2$
(iii) $a_{ii} \leqslant 0$ for all i, $a_{kk} < 0$ for at least one k
(iv) there exists a nonzero term in the expansion of det A.

Interestingly, the necessity parts of the proofs use the Routh–Hurwitz theorem, whilst sufficiency is established via the Liapunov result (Theorem 4.2).

Other problems in the qualitative theory of matrices which have been studied because of their relevance in mathematical economics include:

(1) Consider the system of linear equations $Ax = b$ where x and b are column n-vectors. This is called *sign solvable* if a knowledge of the sign patterns of A and b determines the sign pattern of x, i.e. if

$$A^{(i)} \in Q(A),\ b^{(i)} \in Q(b),\ A^{(i)}x^{(i)} = b^{(i)} \Rightarrow x^{(i)} \in Q(x).$$

What are necessary and sufficient conditions for sign solvability?

(2) The set $Q(A)$ is said to have *signed inverse* if $B \in Q(A) \Rightarrow B^{-1} \in Q(A^{-1})$, i.e. if every matrix which is sign similar to A has an inverse which is sign similar to A^{-1}. What connection does this have with sign stability? More specifically, for what sign pattern does a knowledge of stability lead to a knowledge of the sign pattern of the inverse?

(3) Problems involving nonnegative matrices $A = [a_{ij}]$ with $a_{ij} \geqslant 0$ (we write $A \geqslant 0$).

A fundamental tool in the investigation of many of these problems is the theory of chains and cycles in matrices [46]. If $A = [a_{ij}]$, a *chain* of length r from i_r to i_{r+1} is a product

$$a_{i_1i_2}a_{i_2i_3}\cdots a_{i_{r-1}i_r}a_{i_ri_{r+1}}, \qquad i_1, i_2, \ldots, i_r \text{ distinct.}$$

A *cycle* of length r is the product

$$a_{i_1i_2}a_{i_2i_3}\cdots a_{i_{r-1}i_r}a_{i_ri_1}, \qquad i_1, i_2, \ldots, i_r \text{ distinct.}$$

Questions of type (1) were first considered by Lancaster [34–37] and Gorman [26]. The following theorem gives necessary and sufficient conditions.

† *See Appendix* 1.

THEOREM 4.20 [14] $Ax = b$ is sign solvable if and only if by renumbering of equations and/or variables and by multiplication of equations and/or variables by -1, $Ax = b$ may be transformed into $\hat{A}\hat{x} = \hat{b}$ where $\hat{x} \leqslant 0, \hat{b} \geqslant 0$ and every diagonal element of $\hat{A}$ is negative, $\hat{A}$ satisfying

(i) every cycle of $\hat{A}$ is non-positive
(ii) $\hat{b}_j > 0$ implies every chain of $\hat{A}$ terminating at j is non-negative.

A variation on this problem [61] is when the magnitudes of the elements of A are such that it is a stability matrix. This has been solved in the case when A is indecomposable, has negative principal diagonal and $b_i \neq 0$ for at most one i [14].

Bassett with others [13, 14] has given some interesting theorems concerning question (2). For example,

THEOREM 4.21 [13] Assume A is a sign stable matrix with all $a_{ii} < 0$. Then A^{-1} is of known sign pattern only if for any $a_{ji} \neq 0$, $a_{jk} \neq 0$, $a_{ki} \neq 0$ we have $\operatorname{sgn}(a_{ji}) = \operatorname{sgn}(a_{jk}a_{ki})$ $(i \neq j \neq k)$.

THEOREM 4.22 [14] Assume A is indecomposable, $a_{ii} < 0$ (all i) and $Q(A)$ has signed inverse. Then A is sign stable if and only if A is *combinatorially symmetric* (i.e. $a_{ij} \neq 0 \Rightarrow a_{ji} \neq 0$, all i, j).

Basic material on nonnegative matrices may be found in [17], [24] and [45]. In economics the following definition has been found useful [50]:

A is a *Morishima matrix* if a permutation matrix P exists such that

$$P^T A P = \begin{bmatrix} A_{11} & A_{12} \\ A_{21} & A_{22} \end{bmatrix},$$

where A_{11}, A_{22} are nonnegative square matrices and $-A_{12}$, $-A_{21}$ are nonnegative rectangular matrices. Nonnegative matrices are thus a special case of Morishima matrices (when A_{22} has zero dimensions). Examples of the way in which Morishima matrices are intimately involved with the other problems are provided by the following two results. First define $Q^*(A) = \{B \mid B$ is sign similar to A and B is a stability matrix$\}$. Then

THEOREM 4.23 [13] Assume A has $a_{ii} < 0$, all i and $a_{ij} \neq 0$, all i, j. Then $B, C \in Q^*(A)$ implies $B^{-1} \in Q(C^{-1})$ if and only if (i) $n = 2$ or (ii) $A = M - aI_n$, where M is a Morishima matrix and $a > m_{ii}$, all i.

Next, define A to be a *quasi-dominant diagonal* matrix if there exist positive scalars $k_1, k_2, \ldots, k_n$ such that

$$k_i|a_{ii}| > \sum_{j \neq i} k_j|a_{ij}|, \qquad (i = 1, 2, \ldots, n), \tag{4.3.1}$$

or if (4.3.1) applies to A^T. (If all the k_i are unity then (4.3.1) reduces to the usual definition of diagonal dominance.) Then

THEOREM 4.24 [13] Let $A = M - aI_n$, where M is an indecomposable Morishima matrix and $a > 0$. Then A is a stability matrix if and only if A is a quasi-dominant diagonal matrix with all $a_{ii} < 0$.

There are many other interesting interrelationships between the various problem areas outlined above and we have only tried to give the reader the flavour of the sort of results which have been obtained. For additional details, references and suggestions for possible future research the original sources should be consulted.

EXERCISES

4.3.1 [71] Show that the matrix $S - kI_n$, where S is a real skew symmetric matrix and k is a real positive scalar, is D-stable.

4.3.2 [56] If A is sign stable show that sgn (det A) $= (-1)^n$. Hence show that if A_{ij} is the cofactor of a_{ij} then
sgn $(A_{ij}) = (-1)^n$ sgn (a_{ij}), provided $a_{ij} \neq 0$, $A_{ij} \neq 0$.

4.3.3 [14] If $Ax = b$ is sign solvable show that det $B \neq 0$ for any $B \in Q(A)$.

4.3.4 [44]. If A is quasi-dominant diagonal and all $a_{ii} < 0$, show that if Re $(\lambda) \geqslant 0$, $\lambda I_n - A$ is also quasi-dominant diagonal.

It can be shown that a quasi-dominant diagonal matrix is nonsingular. Hence deduce that under the above assumptions A is a stability matrix.

(In fact a converse result also holds [44].)

4.3.5 Use Gershgorin's theorem [p. 203] to show that if A is real and diagonal dominant with all $a_{ii} < 0$ then it is a stability matrix.

Hence show that any matrix having all elements on the principal diagonal negative is potentially stable.

REFERENCES

1. ANDERSON, B. D. O., 'A system theory criterion for positive real matrices', *SIAM J. Control*, **5**, 171–182 (1967).
2. ANDERSON, B. D. O., 'Application of the second method of Lyapunov to the proof of the Markov stability criterion', *Int. J. Control*, **5**, 473–482 (1967).

3. BALLANTINE, C. S., 'A note on the matrix equation $H = AP+PA^*$', *Linear Algebra and its Applications*, **2**, 37–47 (1969).
4. BARNETT, S. and STOREY, C., 'Stability analysis of constant linear systems by Lyapunov's second method', *Electronics Letters*, **2**, 165–166 (1966).
5. BARNETT, S. and STOREY, C., 'Analysis and synthesis of stability matrices', *J. Diff. Eqns.* **3**, 414–422 (1967).
6. BARNETT, S. and STOREY, C., 'On the general-functional matrix for a linear system', *IEEE Trans. Aut. Control*, **AC-12**, 436–438 (1967).
7. BARNETT, S. and STOREY, C., 'Comment on "Invariance and sensitivity"', *IEEE Trans. Aut. Control*, **AC-12**, 117–118 (1967).
8. BARNETT, S. and STOREY, C., 'Some results on the sensitivity and synthesis of asymptotically stable linear and nonlinear systems', *Automatica*, **4**, 187–194 (1968).
9. BARNETT, S. and STOREY, C., 'Some applications of the Lyapunov matrix equation', *J. Inst. Maths. Applics.* **4**, 33–42 (1968).
10. BARNETT, S., 'Algebraic methods in Lyapunov theory', *Control*, **12**, 545–549 (1968).
11. BARNETT, S., FORD, B., and STOREY, C., 'Comparison of several methods for calculating the general functional matrix for linear systems', UKAC Conference, Leicester (1968). Also, *Int. J. Control*, **10**, 677–686 (1969).
12. BARNETT, S. and STOREY, C., *Matrix methods in stability theory*, Nelson, London (1970).
13. BASSETT, L., HABIBAGAHI, H., and QUIRK, J., 'Qualitative economics and Morishima matrices', *Econometrica*, **35**, 221–233 (1965).
14. BASSETT, L., MAYBEE, J., and QUIRK, J., 'Qualitative economics and the scope of the correspondence principle', *Econometrica*, **36**, 544–563 (1968).
15. BELLMAN, R., *Introduction to matrix analysis*, 2nd Edn., McGraw-Hill, New York (1970).
16. BICKLEY, W. G. and MACNAMEE, J., 'Matrix and other direct methods for the solution of systems of linear difference equations', *Phil. Trans. Roy. Soc. Lond.* Series A, **252**, 69–131 (1960).
17. BRAUER, A., 'On the characteristic roots of non-negative matrices', in *Recent advances in matrix theory*, Ed. H. Schneider, University of Wisconsin Press (1964).
18. CARLSON, D. and SCHNEIDER, H., 'Inertia theorems for matrices: the semidefinite case', *J. Math. Anal. Appl.* **6**, 430–446 (1963).
19. CARLSON, D., 'Rank and inertia bounds for matrices under $R(AH) \geqslant 0$', *J. Math. Anal. Appl.* **10**, 100–111 (1965).

20. CARLSON, D., 'On real eigenvalues of complex matrices', *Pacific J. Math.* **15,** 1119–1129 (1965).
21. CARLSON, D., 'A new criterion for H-stability of complex matrices', *Linear Algebra and its Applications,* **1,** 59–64 (1968).
22. DAVISON, E. J., 'On pole assignment in linear systems with incomplete state feedback', *IEEE Trans. Aut. Control,* **AC-15,** 348–351 (1970).
23. FULLER, A. T., 'Conditions for a matrix to have only characteristic roots with negative real parts', *J. Math. Anal. Appl.* **23,** 71–98 (1968).
24. GANTMACHER, F. R., *Theory of matrices,* Vols. I, II, Chelsea Publishing Company, New York (1960).
25. GIVENS, W., 'Elementary divisors and some properties of the Lyapunov mapping $X \rightarrow AX + XA^*$', *Argonne Nat. Lab. Report ANL-6456* (1961).
26. GORMAN, T., 'More scope for qualitative economics', *Review of Economic Studies,* **31,** 65–68 (1964).
27. HAYNSWORTH, E. V., 'Determination of the inertia of a partitioned Hermitian matrix', *Linear Algebra and its Applications,* **1,** 73–81 (1968).
28. HAYNSWORTH, E. V. and OSTROWSKI, A. M., 'On the inertia of some classes of partitioned matrices', *Linear Algebra and its Applications,* **1,** 299–316 (1968).
29. HEYMANN, M., 'Comments on "Pole assignment in multi-input controllable linear systems"', *IEEE Trans. Aut. Control,* **AC-13,** 748–749 (1968).
30. HILL, R. D., 'Inertia theory for simultaneously triangulable complex matrices', *Linear Algebra and its Applications,* **2,** 131–142 (1969).
31. JOYCE, G. T., and BARNETT, S., 'Remarks on the inertia of a matrix', *Linear Algebra and its Applications,* **3,** 1–5 (1970).
32. KALMAN, R. E. and BERTRAM, J. E., 'Control system analysis and design via the second method of Liapunov', *Trans. ASME, J. Basic Eng.* **82D,** 371–393 (1960).
33. KALMAN, R. E., 'Lyapunov functions for the problem of Lur'e in automatic control', *Proc. Nat. Acad. Sci. USA,* **49,** 201–205 (1963).
34. LANCASTER, K., 'The scope of qualitative economics', *Review of Economic Studies,* **29,** 99–123 (1962).
35. LANCASTER, K., 'Partitionable systems and qualitative economics', *Review of Economic Studies,* **31,** 69–72 (1964).
36. LANCASTER, K., 'The theory of qualitative linear systems', *Econometrica,* **33,** 395–408 (1965).

37. LANCASTER, K., 'The solution of qualitative comparative static problems', *Quart. Jnl. Economics*, **53**, 278–295 (1966).
38. LEHNIGK, S. H., *Stability theorems for linear motions with an introduction to Liapunov's direct method*, Prentice-Hall, Englewood Cliffs, New Jersey (1966).
39. LEHNIGK, S. H., 'Liapunov's direct method and the number of zeros with positive real parts of a polynomial with constant complex coefficients', *SIAM J. Control*, **5**, 234–244 (1967).
40. MACFARLANE, A. G. J., 'The calculation of functionals of the time and frequency response of a linear constant coefficient dynamical system', *Quart. Jnl. Mech. and Appl. Math.* **16**, 259–271 (1963).
41. MACFARLANE, A. G. J., 'Functional-matrix theory for the general linear network', *Proc. IEE*, **112**, 763–770 (1965); **113**, 1268–1276 (1966).
42. MACFARLANE, A. G. J. and SABOUNI, R., 'Functional-matrix theory for the general linear electrical network. Part 4. Bounds on eigenvalues and network reduction', *Proc. IEE*, **115**, 755–758 (1968).
43. MACFARLANE, A. G. J., 'Functional-matrix theory for the general linear electrical network. Part 5. Use of Kronecker products and Kronecker sums', *Proc. IEE*, **116**, 1745–1747 (1969).
44. MCKENZIE, L., 'Matrices with dominant diagonals and economic theory', in *Mathematical methods in the social sciences*, Ed. K. J. Arrow, S. Karlin and P. Suppes, Stanford University Press, Stanford, California (1960).
45. MARCUS, M. and MINC, H., *A survey of matrix theory and matrix inequalities*, Allyn and Bacon, Boston (1964).
46. MAYBEE, J. S., 'Remarks on the theory of cycles in matrices', *Department of Statistics Report*, Purdue University (1965).
47. MAYBEE, J. S. and QUIRK, J. P., 'Qualitative problems in matrix theory', *SIAM Review*, **11**, 30–51 (1969).
48. METZLER, L. A., 'Stability of multiple markets: the Hicks conditions', *Econometrica*, **13**, 277–292 (1945).
49. MOLINARI, B. P., 'Algebraic solution of matrix linear equations in control theory', *Proc. IEE*, **116**, 1748–1754 (1969).
50. MORISHIMA, M., 'On the laws of change of the price system in an economy which contains complementary commodities', *Osaka Economic Papers*, **1**, 101–113 (1952).
51. OSTROWSKI, A. M., 'A quantitative formulation of Sylvester's law of inertia', *Proc. Nat. Acad. Sci. USA*, **45**, 740–744 (1959).
52. OSTROWSKI, A. M. and SCHNEIDER, H., 'Some theorems on the inertia of general matrices', *J. Math. Anal. Appl.* **4**, 72–84 (1962).

53. PARKS, P. C., 'A new proof of the Routh–Hurwitz stability criterion using the second method of Liapunov', *Proc. Cambridge Philos. Soc.* **58,** 694–702 (1962).
54. PARKS, P. C., 'Hermite–Hurwitz and Hermite–Bilharz links using matrix multiplication', *Electronics Letters,* **5,** 55–57 (1969).
55. POWER, H. M., 'Canonical form for the matrices of linear discrete-time systems', *Proc. IEE,* **116,** 1245–1252 (1969).
56. QUIRK, J. and RUPPERT, R., 'Qualitative economics and the stability of equilibrium', *Review of Economic Studies,* **32,** 311–326 (1965).
57. REDHEFFER, R., 'Remarks on a paper by Taussky', *J. Algebra,* **2,** 42–47 (1965).
58. REIS, G. C., 'Some properties of indefinite matrices related to control theory', *IEEE Trans. Aut. Control,* **AC-12,** 789–790 (1967).
59. REZA, F., 'Derived stable matrices', *Proc. IEEE,* **55,** 1112 (1967).
60. ROSENBROCK, H. H., 'On the design of linear multivariable control systems', *IFAC Conference,* London (1966).
61. SAMUELSON, P., *The foundations of economic analysis,* Harvard University Press, Cambridge, Massachusetts (1955).
62. SCHULTZ, D. G., 'The generation of Liapunov functions', in *Advances in Control Systems,* Vol. 2, Academic Press, New York (1965).
63. SCHULTZ, D. G. and MELSA, J. L., *State functions and linear control systems,* McGraw-Hill, New York (1967).
64. SMITH, R. A., 'Bounds for quadratic Lyapunov functions', *J. Math. Anal. Appl.* **12,** 425–435 (1965).
65. SMITH, R. A., 'Matrix calculations for Liapunov quadratic forms', *J. Diff. Eqns.* **2,** 208–217 (1966).
66. SMITH, R. A., 'Sufficient conditions for stability of a solution of difference equations', *Duke Math. J.* **33,** 725–734 (1966).
67. SMITH, R. A., 'On minimising the condition of a Lyapunov matrix', *J. London Math. Soc.* **43,** 675–678 (1968).
68. STEIN, P., 'Some general theorems on iterants', *J. Res. NBS,* **48,** 82–83 (1952).
69. STEIN, P., 'On the ranges of two functions of positive definite matrices', *J. Algebra,* **2,** 350–353 (1965).
70. STEIN, P. and PFEFFER, A., 'On the ranges of two functions of positive definite matrices II', *ICC Bulletin,* **6,** 81–86 (1967).
71. TAUSSKY, O., 'A remark on a theorem of Lyapunov', *J. Math. Anal. Appl.* **2,** 105–107 (1961).
72. TAUSSKY, O., 'A generalization of a theorem of Lyapunov', *SIAM J. Appl. Math.* **9,** 640–643 (1961).

73. TAUSSKY, O. and WIELANDT, H., 'On the matrix function $AX + X'A'$', *Arch. Rat. Mech. and Anal.* **9,** 93–96 (1962).
74. TAUSSKY, O., 'Matrices C with $C^n \to 0$', *J. Algebra,* **1,** 5–10 (1964).
75. TAUSSKY, O., 'On the variation of the characteristic roots of a finite matrix under various changes of its elements', in *Recent Advances in Matrix Theory,* Ed. H. Schneider, University of Wisconsin Press (1964).
76. TAUSSKY, O., 'Stable matrices', in *Programmation en Mathématiques Numériques,* Centre National de la Recherche Scientifique, Paris (1968).
77. TRAMPUS, A., 'A canonical basis for the matrix transformation $X \to AX + XB$', *J. Math. Anal. Appl.* **14,** 242–252 (1966).
78. VOGT, W. G., 'Discussion of "A modified Lyapunov method for non-linear stability analysis"', *IRE Trans.* **AC-7,** 85–88, January (1962), 90–92, April (1962).
79. VOGT, W. G., 'Transient response from the Lyapunov stability equation', *Preprints, Joint Automatic Control Conference,* Troy, New York (1965).
80. WEAVER, L. E., *Reactor dynamics and control,* Elsevier, New York (1968).
81. WIGNER, E. P. and YANASE, M. M., 'On the positive semidefinite nature of a certain matrix expression', *Canad. J. Math.* **16,** 397–406 (1964).
82. WILKINSON, J. H., *The algebraic eigenvalue problem,* Oxford University Press (1965).
83. WONHAM, W. M., 'On pole assignment in multi-input controllable linear systems', *IEEE Trans. Aut. Control,* **AC-12,** 660–665 (1967).

5

Matrix Riccati Equations

5.1 General matrix Riccati differential equations

The most general equation we shall be discussing can be written in the form

$$\dot{W}(t) = W(t)E(t)W(t) + D(t)W(t) + W(t)F(t) + G(t), \quad (5.1.1)$$

where $D(t)$, $E(t)$, $F(t)$ and $G(t)$ are given $n \times n$ matrices the elements of which are real continuous functions of the real variable t on some interval $t_0 \leqslant t \leqslant t_1$. The equation (5.1.1) is called a differential equation of *Riccati type*, or simply a *Riccati equation*, by analogy with the well-known scalar equation (see, for example, [29]):

$$\dot{w} = ew^2 + fw + g, \quad (5.1.2)$$

where e, f and g are scalar functions of t. The important feature which both equations have in common is, of course, the quadratic term in the unknown scalar or matrix function, and in the case of (5.1.2) this is easily dealt with by means of a simple transformation which reduces the equation to one which is linear (although of second order—see Exercise 5.1.1).

Equations of the type (5.1.1) have been studied by mathematicians for a number of years, but it is only quite recently that a particular case of (5.1.1) has attracted widespread attention. This is because of the fundamental way in which a matrix Riccati equation arises from an optimization problem in linear control system theory. This will be discussed in detail in the next two sections of this chapter, and we shall be concerned here with properties of general matrix Riccati differential equations.

We first consider a *linear* matrix differential equation which is important in the subsequent development. This relies on the well-known result (see, for example, [9], p. 171) that the solution of

$$\dot{X}(t) = A(t)X(t), \qquad X(t_0) = C,$$

where all matrices are $n \times n$, is unique and nonsingular. It is then easy to derive:

THEOREM 5.1 [32] For every constant $n \times n$ matrix X_0, the equation

$$\dot{X}(t) = H(t)X(t) + X(t)K(t), \quad X(t_0) = X_0, \tag{5.1.3}$$

where $H(t)$ and $K(t)$ are $n \times n$ matrices continuous on $t_0 \leqslant t \leqslant t_1$, has a unique solution $X(t) = X_1(t)X_0X_2(t)$, where $X_1(t)$ and $X_2(t)$ are the solutions of

$$\left.\begin{aligned} \dot{X}_1(t) &= H(t)X_1(t), \qquad X_1(t_0) = I_n \\ \dot{X}_2(t) &= X_2(t)K(t), \qquad X_2(t_0) = I_n. \end{aligned}\right\} \tag{5.1.4}$$

Furthermore, rank $X(t) =$ rank X_0, $t_0 \leqslant t \leqslant t_1$.

This result enables us to relate (5.1.1) to a corresponding homogeneous equation, as follows:

THEOREM 5.2 [32] If $W_1(t)$ and $W_2(t)$ are two solutions of (5.1.1), continuous on $t_0 \leqslant t \leqslant t_1$, then $V(t) = W_2(t) - W_1(t)$ is a solution of

$$\dot{V}(t) = [D(t) + W_1(t)E(t)]V(t) + V(t)[F(t) + E(t)W_1(t)] + V(t)E(t)V(t), \tag{5.1.5}$$

and $V(t)$ is continuous and has constant rank on $t_0 \leqslant t \leqslant t_1$.

Proof. It is easy to verify that $W_2 - W_1$ satisfies equation (5.1.5). Also, this equation can be written in the form (5.1.3), with

$$H(t) = D(t) + W_1(t)E(t) + V(t)E(t)$$
$$K(t) = F(t) + E(t)W_1(t),$$

so by Theorem 5.1, rank $V(t) =$ rank $V(t_0) =$ rank $[W_2(t_0) - W_1(t_0)]$ and is therefore constant.

Notice that (5.1.5) is a *homogeneous* Riccati equation. We shall be giving an explicit solution of a similar homogeneous equation in terms of the solution of the following inhomogeneous version of (5.1.3):

$$\dot{X}(t) = H(t)X(t) + X(t)K(t) + L(t). \tag{5.1.6}$$

THEOREM 5.3 [30] The solution of (5.1.6) satisfying $X(t_0) = X_0$ is unique and is given by

$$X(t) = X_1(t)X_0X_2(t) + X_1(t)\left\{\int_{t_0}^{t} X_1^{-1}(s)L(s)X_2^{-1}(s)\, ds\right\}X_2(t), \tag{5.1.7}$$

the matrices $X_1(t)$ and $X_2(t)$ being determined by (5.1.4).

Proof. Straightforward differentiation of (5.1.7) establishes that it satisfies (5.1.6). Clearly the desired initial condition is also satisfied, so it remains to prove uniqueness. If $Y(t)$ is a second solution of (5.1.6), then $Z(t) = X(t) - Y(t)$ satisfies (5.1.3) with $Z(t_0) = 0$, and hence by Theorem 5.1, $Z(t) \equiv 0$ as required.

Notice that when $H(t) = K^T(t)$ then it follows from (5.1.4) that $X_2^T(t) = X_1(t)$, and hence from (5.1.7) that the corresponding solution of (5.1.6) will be symmetric if X_0 and $L(t)$ are symmetric. In fact we can state a stronger result:

THEOREM 5.4 [30] If in (5.1.6), $H(t) = K^T(t)$ and if $L(t)$ and X_0 are symmetric then the solution $X(t)$ in (5.1.7) is positive definite† symmetric if and only if

$$X_0 + \int_{t_0}^{t} X_1^{-1}(s)L(s)\{X_1^T(s)\}^{-1}\, ds$$

is positive definite symmetric for $t_0 \leqslant t \leqslant t_1$.

The preceding theorem gives us a class of equations of the form (5.1.6) which have a nonsingular solution, and provides a link with homogeneous Riccati equations:

THEOREM 5.5 [32] If the solution $X(t)$ of (5.1.6) is nonsingular on $t_0 \leqslant t \leqslant t_1$, then $U(t) = X^{-1}(t)$ is the unique solution of

$$-\dot{U}(t) = U(t)L(t)U(t) + K(t)U(t) + U(t)H(t) \qquad (5.1.8)$$

satisfying

$$U(t_0) = U_0 = X_0^{-1}.$$

Proof. This follows at once by differentiation of $X^{-1}(t)$.

We thus have an explicit solution of the homogeneous Riccati equation (5.1.8) in terms of the solution (5.1.7) of a linear matrix differential equation.

We now return to the general Riccati equation (5.1.1) and show how this can be related to a set of $2n$ linear differential equations (see Exercise 5.1.2 for the analogous scalar case). Consider the equations

$$\left.\begin{aligned} \dot{\phi} &= -F(t)\phi(t) - E(t)\psi(t) \\ \dot{\psi} &= G(t)\phi(t) + D(t)\psi(t) \end{aligned}\right\} \qquad (5.1.9)$$

where $\phi(t)$, $\psi(t)$ are column n-vectors with components $\phi_i(t)$, $\psi_i(t)$ respectively. A fundamental relationship between (5.1.1) and (5.1.9) is given by

THEOREM 5.6 [32] There exists a set of n solutions

$$\phi^{(j)} = [\phi_{1j}, \ldots, \phi_{nj}]^T, \psi^{(j)} = [\psi_{1j}, \ldots, \psi_{nj}]^T, (j = 1, 2, \ldots, n) \quad (5.1.10)$$

† *See Appendix* 3.

of (5.1.9) such that the matrix $Y(t) = [\phi_{ij}(t)]$, $i, j = 1, \ldots, n$, is nonsingular on $t_0 \leqslant t \leqslant t_1$ if and only if (5.1.1) possesses a continuous solution on $t_0 \leqslant t \leqslant t_1$. Furthermore, if $Z(t) = [\psi_{ij}(t)]$, $i, j = 1, \ldots, n$, then $W(t) = Z(t)Y^{-1}(t)$ is a solution of (5.1.1) continuous on $t_0 \leqslant t \leqslant t_1$.

Proof. Let $W(t)$ be a continuous solution of (5.1.1) on $t_0 \leqslant t \leqslant t_1$, and consider the linear homogeneous matrix equation

$$\dot{Y}(t) = -[F(t) + E(t)W(t)]Y(t). \tag{5.1.11}$$

Let $Y(t) = [\phi_{ij}(t)]$ be a (nonsingular) solution of (5.1.11), and define the matrix $Z = [\psi_{ij}(t)]$ by $Z(t) = W(t)Y(t)$. From (5.1.11) and (5.1.1) we have

$$\left.\begin{aligned} \dot{Y}(t) &= -F(t)Y(t) - E(t)Z(t) \\ \dot{Z}(t) &= G(t)Y(t) + D(t)Z(t) \end{aligned}\right\} \tag{5.1.12}$$

and this choice for the elements ϕ_{ij}, ψ_{ij} in (5.1.10) gives n solutions of (5.1.9).

Conversely, if (5.1.10) represents a set of solutions of (5.1.9) such that $[\phi_{ij}(t)]$ is nonsingular, then $Y = [\phi_{ij}(t)]$, $Z = [\psi_{ij}(t)]$ satisfy (5.1.12), and it is easy to verify that $W(t) = Z(t)Y^{-1}(t)$ is a solution of (5.1.1) continuous on $t_0 \leqslant t \leqslant t_1$.

The quadratic differential equation (5.1.1) may thus be replaced by the linear matrix equations (5.1.12).

Levin [23] has considered equation (5.1.1) when D, E, F, G and W have dimensions respectively $n_1 \times n_1$, $n_2 \times n_1$, $n_2 \times n_2$, $n_1 \times n_2$ and $n_1 \times n_2$, and obtained a result corresponding to Theorem 5.6. He has also proved the following theorem (when $n_1 = n_2$), which is a generalization of a well-known result for the scalar Riccati equation (5.1.2) [16, p. 23].

THEOREM 5.7 If $W_i(t)$, $i = 1, 2, 3, 4$ are four solutions of (5.1.1) continuous on $t_0 \leqslant t \leqslant t_1$, and if each matrix $W_i(t_0) - W_j(t_0)$ is nonsingular for $i \neq j$, then

$$[W_3(t) - W_1(t)]^{-1}[W_3(t) - W_2(t)][W_4(t) - W_2(t)]^{-1}[W_4(t) - W_1(t)] = B(t)CB^{-1}(t)$$

where $B(t)$ and the constant matrix C are nonsingular.

Proof. If we let $U_i(t) = W_{i+1}(t) - W_1(t)$, then because of the assumptions in the theorem it follows at once from Theorem 5.2 that $U_i(t)$ is nonsingular ($i = 1, 2, 3$; $t_0 \leqslant t \leqslant t_1$) and is the solution of

$$\dot{U}_i(t) = [D(t) + W_1(t)E(t) + U_i(t)E(t)]U_i(t) + U_i(t)[F(t) + E(t)W_1(t)]. \tag{5.1.13}$$

The matrices $U_i(t) - U_j(t)$, $i \neq j$, are also nonsingular. Writing

$V_i(t) = U_i^{-1}(t)$, use of (5.1.13) establishes that

$$\dot{V}_i(t) = -[F(t) + E(t)W_1(t)]V_i(t) - V_i(t)[W_1(t)E(t) + D(t)] - E(t). \quad (5.1.14)$$

If $T_i(t) = V_{i+1}(t) - V_1(t)$, then (5.1.14) leads to

$$\dot{T}_i(t) = -[F(t) + E(t)W_1(t)]T_i(t) - T_i(t)[W_1(t)E(t) + D(t)]. \quad (5.1.15)$$

Also,

$$\begin{aligned} T_i(t) &= U_{i+1}^{-1}(t) - U_1^{-1}(t) \\ &= U_{i+1}^{-1}(t)[U_1(t) - U_{i+1}(t)]U_1^{-1}(t) \end{aligned}$$

which shows that $T_i(t)$ is nonsingular ($i = 1, 2, 3, t_0 \leqslant t \leqslant t_1$). Application of Theorem 5.1 to (5.1.15) then gives us

$$T_i(t) = X_1(t)X_{i0}X_2(t),$$

where $T_i(t_0) = X_{i0}$ and

$$\begin{aligned} \dot{X}_1(t) &= -[F(t) + E(t)W_1(t)]X_1(t), & X_1(t_0) &= I_n \\ \dot{X}_2(t) &= -X_2(t)[W_1(t)E(t) + D(t)], & X_2(t_0) &= I_n. \end{aligned}$$

Therefore,

$$\begin{aligned} [W_{i+2}(t) - W_1(t)]^{-1} - [W_2(t) - W_1(t)]^{-1} &= U_{i+1}^{-1}(t) - U_1^{-1}(t) \\ &= V_{i+1}(t) - V_1(t) \\ &= T_i(t) \\ &= X_1(t)X_{i0}X_2(t). \end{aligned}$$

Multiplying this last expression by $W_2 - W_1$ on the right and $W_{i+2} - W_1$ on the left, we obtain

$$W_2(t) - W_{i+2}(t) = [W_{i+2}(t) - W_1(t)]X_1(t)X_{i0}X_2(t)[W_2(t) - W_1(t)]. \quad (5.1.16)$$

Since $W_2 - W_3$ and $W_2 - W_4$ are nonsingular, setting $i = 1, 2$ in (5.1.16) and equating the resulting expressions gives

$$\begin{aligned} &(W_2 - W_3)^{-1}(W_3 - W_1)X_1X_{10}X_2(W_2 - W_1) \\ &\quad = (W_2 - W_4)^{-1}(W_4 - W_1)X_1X_{20}X_2(W_2 - W_1). \quad (5.1.17) \end{aligned}$$

Since $W_2 - W_1$, X_1, X_{20} and X_{10} are all nonsingular, (5.1.17) is easily rearranged to give the desired result with $B(t) = X_1(t)$, $C = X_{10}X_{20}^{-1}$.

The proof which we have just given closely follows that for the scalar case [16]. Using a similar argument Reid [32] has obtained a different theorem involving $n^2 + 3$ solutions of (5.1.1), which also reduces to the classical result when $n = 1$. The same author in a later paper [33] has given yet another different generalization of Theorem 5.7.

EXERCISES

5.1.1 Show that the substitution $w = -\dot{y}/ey$, where y is a scalar function of t, reduces (5.1.2) to a second order linear equation in y.

5.1.2 Show that $w(t) = y_2(t)/y_1(t)$, where y_1 and y_2 are solutions of

$$\dot{y}_1 = -\tfrac{1}{2}fy_1 - ey_2$$
$$\dot{y}_2 = gy_1 + \tfrac{1}{2}fy_2,$$

is a solution of (5.1.2).

5.1.3 Verify that (5.1.3) is satisfied by the given solution.

5.1.4 [32] Use Theorem 5.1 to obtain the general solution of

$$\dot{X}(t) = H(t)X(t) - X(t)H(t).$$

5.1.5 [10] Show that the substitution $Y(t) = \dot{X}(t)X^{-1}(t)$ in the equation $\ddot{X}(t) - K(t)X(t) = 0$ reduces it to a simple case of (5.1.1).

5.1.6 [32] If $V_1(t)$ and $V_2(t)$ are nonsingular continuous solutions of

$$\dot{V}(t) = V(t)E(t)V(t) + D(t)V(t) + V(t)F(t) \text{ on } t_0 \leqslant t \leqslant t_1,$$

show that $V_2^{-1}(t) - V_1^{-1}(t)$ has constant rank on this interval.

5.1.7 Let $Y(t)$ and $Z(t)$ be solutions of $\dot{Y}(t) = A(t)Y(t)$ and $\dot{Z}(t) = -Z(t)A(t)$ respectively (all matrices are $n \times n$). Show that if $W = YZ$ then trace $W(t)$ is constant.

See [38] for a more general result along these lines.

5.1.8 In equation (5.1.3) let $H(t) = A^T$ and $K(t) = A$, where A is a constant $n \times n$ stability matrix. Integrate the resulting equation and use Theorem 5.1 to show that the solution of

$$A^TY + YA = -X_0,$$

where Y is a constant matrix, is

$$Y = \int_{t_0}^{\infty} \exp(A^Tt)X_0 \exp(At)\, dt.$$

We recognize the equation for Y as a Liapunov matrix equation (see Chapter 4, Section 1).

5.1.9 Show, by considering the matrix $Y^T(t)Z(t)$ in Theorem 5.6, that if $F = D^T$, $E = E^T$, $G = G^T$ and $W(t_s)$ is symmetric for some specified time t_s, then the solution $W(t)$ of equation (5.1.1) is symmetric for all t.

5.2 Riccati differential equation in optimal control

We now concentrate on the important special case of equation (5.1.1) which arises in the theory of optimal linear control systems. Specifically, let the equations describing the system be

$$\left.\begin{aligned} \dot{x}(t) &= A(t)x(t) + B(t)u(t) \\ x(t_0) &= x_0, \end{aligned}\right\} \tag{5.2.1}$$

where, as in previous chapters, x is a column n-vector representing the state of the system, u is a column m-vector of controlled variables and $A(t)$, $B(t)$ are respectively $n \times n$ and $n \times m$, but now having elements which are continuous functions of time. The problem is to find a vector function $u(t)$ which minimizes (subject to (5.2.1)) the functional

$$\mathcal{J} = \tfrac{1}{2}x^T(t_1)Mx(t_1) + \tfrac{1}{2}\int_{t_0}^{t_1} [x^T(t)Q(t)x(t) + u^T(t)R(t)u(t)]\, dt, \qquad (5.2.2)$$

where M and $Q(t)$ are symmetric and positive semidefinite and $R(t)$ is symmetric positive definite. Since the terms in (5.2.2) are all quadratic forms, and since $\mathcal{J}$ can be thought of as a measure of the way in which the system behaves, (5.2.2) is usually called a *quadratic performance index*.

The formulation we have just given is known as the optimal state regulator problem, since the aim is to control (or 'regulate') the state vector $x(t)$ so that its elements are kept near to zero without unduly large control effort. An alternative is the output regulator problem, when it is some output vector $y(t) = C(t)x(t)$ which is to be kept near zero, in which case the terms involving $x(t)$ in (5.2.2) are replaced by $y^T(t_1)My(t_1)$ and $y^T(t)Q(t)y(t)$ respectively. Even more generally, the objective may be to make the output as close as possible to some desired output $z(t)$. Setting $e(t) = z(t) - y(t)$, we then replace the relevant terms in (5.2.2) by $e^T(t_1)Me(t_1)$ and $e^T(t)Q(t)e(t)$, so that minimizing the corresponding $\mathcal{J}$ will minimize $\|e(t)\|$. In each case the purpose of the quadratic form in $u(t)$ is to ensure, by suitable choice of the elements of $R(t)$, that in satisfying the objective the control variables $u_i(t)$ are not required to be impractically large.

It is well known (see Appendix 4) that the solution to the problem described by (5.2.1) and (5.2.2) is given by the linear feedback control

$$u = -R^{-1}(t)B^T(t)P(t)x(t), \qquad (5.2.3)$$

where $P(t)$ is the symmetric matrix which satisfies

$$\dot{P}(t) = P(t)B(t)R^{-1}(t)B^T(t)P(t) - A^T(t)P(t) - P(t)A(t) - Q(t) \qquad (5.2.4)$$

subject to $P(t_1) = M$. Clearly (5.2.4) is a special case of (5.1.1) and represents $\tfrac{1}{2}n(n+1)$ equations for the $\tfrac{1}{2}n(n+1)$ elements of $P(t)$. The symmetry of the solution follows from Exercise 5.1.9. Notice that even if A, B, Q and R are all independent of t, the matrix P will still be a function of time, so that the optimal control (5.2.3), although linear in x, will be time-varying.

It should be admitted that the widespread use of a *quadratic* performance index is due to the relative ease with which it can be handled mathematically and to the fact that it results in *linear* feedback. Indeed there is at present no straightforward procedure for choosing the elements of $Q(t)$ and $R(t)$ in advance so as to ensure a satisfactory optimal system, so that as a design method this approach still relies to a certain extent on trial and error. In fact it has been suggested that other performance indices (involving, for example, higher degree terms in the state variables [8, 4] or time weighting factors [27]), whilst leading to more complicated analysis, may nevertheless produce a resulting optimal system which has more desirable properties.

Detailed discussion of the control theory aspects of the problem can be found in a standard text such as [5] and will not be gone into any further here. Note, however, that the theory of optimal linear filtering [19] also depends fundamentally on the solution of a Riccati equation similar to (5.2.4).

If we wish to convert solution of (5.2.4) into an initial value problem this is easily done by setting $\tau = t_0 + t_1 - t$, which transforms (5.2.4) and its associated end condition into

$$-dP(\tau)/d\tau = P(\tau)B(\tau)R^{-1}(\tau)B^T(\tau)P(\tau) - A^T(\tau)P(t) - P(\tau)A(\tau) - Q(\tau) \tag{5.2.5}$$

$$P(t_0) = M, \qquad t_0 \leqslant \tau \leqslant t_1.$$

In the particular case when $Q = 0$, comparing equations (5.2.5) and (5.1.8) gives us $L = BR^{-1}B^T$, $K = -A^T$, $H = -A$, so that the conditions of Theorem 5.4 are satisfied. Hence by Theorem 5.5 the unique solution of (5.2.5) in this case is positive definite and is given by $P(\tau) = X^{-1}(\tau)$, $P(t_0) = M$ where

$$dX/d\tau = -A(\tau)X(\tau) - X(\tau)A^T(\tau) + B(\tau)R^{-1}(\tau)B^T(\tau),$$
$$X(t_0) = M^{-1}, \qquad t_0 \leqslant \tau \leqslant t_1.$$

We also have an important uniqueness result for the inhomogeneous equation:

THEOREM 5.8 [18] There exists a unique symmetric solution $\Pi(t)$, defined on $t_0 \leqslant t \leqslant t_1$, satisfying equation (5.2.4) subject to $\Pi(t_1) = M$. Further, the minimum value of (5.2.2) is $\frac{1}{2}x_0^T\Pi(t_0)x_0$.

The linear equations corresponding to (5.1.9) can be written

$$\left.\begin{aligned} \begin{bmatrix} \dot{x}(t) \\ \dot{p}(t) \end{bmatrix} &= H(t)\begin{bmatrix} x(t) \\ p(t) \end{bmatrix}, \\ \text{where} \qquad H(t) &= \begin{bmatrix} A(t) & -B(t)R^{-1}(t)B^T(t) \\ -Q(t) & -A^T(t) \end{bmatrix} \end{aligned}\right\} \tag{5.2.6}$$

and the column n-vector $p(t)$ is termed the *adjoint vector*. The form of (5.2.6) arises directly when the calculus of variations is applied to the optimization problem stated at the beginning of this section (see Appendix 4). The adjoint and state vectors are related by

$$p(t) = P(t)x(t), \tag{5.2.7}$$

and it is easily verified that substitution of (5.2.7) into (5.2.6) gives the

Riccati equation (5.2.4). This formulation enables us to give an explicit solution of (5.2.4) as follows:

Let the $2n \times 2n$ matrix

$$\Phi(t) = \begin{bmatrix} \phi_1(t) & \phi_2(t) \\ \phi_3(t) & \phi_4(t) \end{bmatrix}$$

(where each block is $n \times n$) be the *transition matrix* of (5.2.6). That is, $\Phi(t)$ is the solution of

$$\frac{d\Phi(t)}{dt} = H(t)\Phi(t), \quad \Phi(t_1) = I_{2n}. \tag{5.2.8}$$

Then we have

THEOREM 5.9 The solution of (5.2.4) satisfying $P(t_1) = M$ is

$$P(t) = [\phi_3(t) + \phi_4(t)M][\phi_1(t) + \phi_2(t)M]^{-1}. \tag{5.2.9}$$

Proof. From Theorem 5.6, the solution of (5.2.4) is $P(t) = Z(t)Y^{-1}(t)$, where

$$\left.\begin{aligned} \dot{Y}(t) &= AY - BR^{-1}B^TZ \\ \dot{Z}(t) &= -QY - A^TZ. \end{aligned}\right\} \tag{5.2.10}$$

Comparison of (5.2.6) and (5.2.10) shows that

$$\begin{bmatrix} Y(t) \\ Z(t) \end{bmatrix} = \Phi(t)\begin{bmatrix} Y(t_1) \\ Z(t_1) \end{bmatrix},$$

which on using the partitioned form of Φ gives

$$Y(t) = \phi_1(t)Y(t_1) + \phi_2(t)Z(t_1), \qquad Z(t) = \phi_3(t)Y(t_1) + \phi_4(t)Z(t_1).$$

Substitution into the expression for $P(t)$ with $M = Z(t_1)Y^{-1}(t_1)$ gives (5.2.9).

It turns out that the matrices $H(t)$ and $\Phi(t)$ have some further interesting algebraic properties which the reader may investigate via Exercises 5.2.5–5.2.9.

Discussion of the relative merits of numerical procedures for solving the Riccati equation (5.2.4) using Theorem 5.9 (or by standard techniques for sets of differential equations) is outside the scope of this book, and we refer the reader to [11], [20], and [37], amongst others, for details.

A rather novel approach has been suggested recently [36] for the case when $m = 1$, $b(t) = [0, 0, \ldots, 0, 1]^T$ and $A(t)$ has the companion type form

$$\begin{bmatrix} 0 & 1 & 0 & \cdots & . \\ 0 & 0 & 1 & \cdots & . \\ . & . & . & \cdots & . \\ . & . & . & \cdots & 1 \\ -a_1(t) & -a_2(t) & -a_3(t) & \cdots & -a_n(t) \end{bmatrix}. \tag{5.2.11}$$

In the performance index (5.2.2), $R(t)$ is taken as unity so the second term in the integrand is just $u^2(t)$. The Riccati equation (5.2.4) to be solved is

$$\left.\begin{aligned}\dot{P}(t) &= P(t)bb^TP(t) - A^T(t)P(t) - P(t)A(t) - Q(t)\\ P(t_1) &= M,\end{aligned}\right\} \tag{5.2.12}$$

and the corresponding optimal closed-loop system is

$$\dot{x}(t) = [A(t) - bb^TP(t)]x(t). \tag{5.2.13}$$

The essence of the method is to reduce (5.2.12) to a much simpler equation in which $A(t)$ is replaced by $\hat{A}$, obtained by setting all $a_i(t) = 0$ in (5.2.11), $Q(t)$ is replaced by a diagonal matrix $\hat{Q}(t)$ and M by $\hat{M}$. The matrices $\hat{Q}(t)$ and $\hat{M}$ are easily obtained from explicit formulae involving the $a_i(t)$ and the elements of $Q(t)$ and M. Further, the optimal closed-loop matrix $\hat{A} - bb^T\hat{P}(t)$ is the same as that in (5.2.13). For various reasons this approach seems promising and it would be useful if it could be extended to problems with more than one control variable.

For optimization problems where the boundary condition for (5.2.4) is not of the form $P(t_s)$ equal to a known matrix for some specified time t_s, Friedland [14] has shown how another transformation to give a pair of matrix differential equations can be used to overcome the difficulty (see Exercise 5.2.3).

A slightly different problem, discussed by Anderson [3], relates to positive-real matrices defined in Chapter 3. Consider the system (5.2.1) with A and B independent of t:

$$\dot{x} = Ax + Bu, \qquad x(t_0) = x_0, \tag{5.2.14}$$

and performance index

$$\int_{t_0}^{t_1} (2x^TSu + u^TRu)\,dt, \tag{5.2.15}$$

where R and S are also independent of t, in place of (5.2.2). Generally, addition of a term $2x^TSu$ into the integrand in (5.2.2) can be handled without difficulty (see Exercise 5.2.4). One of Anderson's results is as follows:

THEOREM 5.10 If (5.2.14) is controllable†, a necessary and sufficient condition for the minimum value of (5.2.15) to be finite valued is that the matrix

$$Z(\lambda) = \tfrac{1}{2}R + S^T(\lambda I - A)^{-1}B$$

be positive-real. The minimum value of (5.2.15) is then $x^T(t_0)\rho(t_0)x(t_0)$ where $\rho(t)$ is the solution of the Riccati equation

$$\left.\begin{aligned}\dot{\rho}(t) &= \rho(t)BR^{-1}B^T\rho(t) - (A^T - SR^{-1}B^T)\rho(t)\\ &\quad - \rho(t)(A - BR^{-1}S^T) + SR^{-1}S^T,\\ \rho(t_1) &= 0.\end{aligned}\right\} \tag{5.2.16}$$

† *See Appendix* 2.

Notice that (5.2.16) can be obtained from (5.2.4) with $A(t)$ replaced by $A - BR^{-1}S^T$, $Q(t)$ replaced by $-SR^{-1}S^T$ and $M = 0$. Further, $[A - BR^{-1}S^T, B]$ is controllable since the pair $[A + BC, B]$ is controllable for any $m \times n$ matrix C if $[A, B]$ is controllable [39].

We close this section by mentioning some work by Bellman [10] for the case when $A = 0$, $B = I_n = R$ and Q is a constant matrix in (5.2.4). By an ingenious technique he obtains bounds for the solution of the corresponding Riccati equation

$$\dot{P}(t) = P^2(t) - Q$$

in the form† $P_2(t) \leqslant P(t) \leqslant P_1(t)$ where $P_1(t)$ and $P_2(t)$ are both solutions of linear matrix differential equations.

EXERCISES

5.2.1 Verify by direct differentiation, using (5.2.8), that the matrix $P(t)$ in (5.2.9) satisfies equation (5.2.4) subject to the condition $P(t_1) = M$.

5.2.2 [6] Show that equation (5.2.4) is unaltered if $A(t)$ is replaced by $A(t) + S(t)P(t)$, where $S(t)$ is an arbitrary skew-symmetric matrix.

Notice that this bears some relationship to Theorem 4.5 (see [7]).

5.2.3 [14] Show that if $P(t)$ is a solution of (5.2.4), so is

$$P_1(t) = P(t) - X(t)Y^{-1}(t)X^T(t),$$

where $X(t)$ and $Y(t)$ satisfy the differential equations

$$\dot{X}(t) = [-A^T(t) + P(t)B(t)R^{-1}(t)B^T(t)]X(t)$$
$$\dot{Y}(t) = X^T(t)B(t)R^{-1}(t)B^T(t)X(t).$$

5.2.4 Show that the form

$$x^TQ_1x + 2x^TSu_1 + u_1^TRu_1$$

can be reduced to the quadratic form in the integrand in (5.2.2) by taking $Q = Q_1 - SR^{-1}S^T$, $u = u_1 + R^{-1}S^Tx$.

5.2.5 [28] If

$$\mathcal{J} = \begin{bmatrix} 0 & -I_n \\ I_n & 0 \end{bmatrix},$$

show that $\mathcal{J}^T = \mathcal{J}^{-1} = -\mathcal{J}$, and that $\mathcal{J}^2 = -I_{2n}$.

A real $2n \times 2n$ matrix H is called *Hamiltonian* if $H = \mathcal{J}H^T\mathcal{J}$. Verify that $H(t)$ in equation (5.2.6) is Hamiltonian.

5.2.6 [28] Show that for *any* Hamiltonian matrix H (defined in the previous exercise), if λ_i is a characteristic root then so is $-\lambda_i$. Show also that if α_i and β_i are right characteristic vectors of H corresponding to λ_i and $-\lambda_i$ respectively, then $-(\mathcal{J}\beta_i)^T$ and $(\mathcal{J}\alpha_i)^T$ are left characteristic vectors corresponding to λ_i and $-\lambda_i$.

† *We use the notation* $X \geqslant Y$ *to mean that* $X - Y$ *is positive semidefinite.*

5.2.7 [28] Let a nonsingular Hamiltonian matrix H have distinct characteristic roots $\pm\lambda_1, \pm\lambda_2, \ldots, \pm\lambda_n$. Using the result of the previous exercise show that

$$L^{-1}HL = \text{diag}\,[\lambda_1, \lambda_2, \ldots, \lambda_n, -\lambda_1, -\lambda_2, \ldots, -\lambda_n],$$

where

$$L = [L_1\, L_2], \qquad L^{-1} = \begin{bmatrix} L_2{}^T\mathcal{J} \\ -L_1{}^T\mathcal{J} \end{bmatrix},$$

$$L_1 = [\alpha_1, \alpha_2, \ldots, \alpha_n], \qquad L_2 = [\beta_1, \beta_2, \ldots, \beta_n].$$

5.2.8 [28] Any real $2n \times 2n$ matrix S is called *symplectic* if $S^T\mathcal{J}S = \mathcal{J}$, where $\mathcal{J}$ is defined in Exercise 5.2.5. Show that the matrix L in Exercise 5.2.7 is symplectic.

5.2.9 [19] Let $\phi(t)$ be the transition matrix for a system $\dot{x} = H(t)x$, where $H(t)$ is an arbitrary Hamiltonian matrix (i.e.

$$\dot{\phi}(t) = H(t)\phi(t),\ \phi(t_0) = I_{2n}).$$

By making the substitution $\psi(t) = [\phi^{-1}(t)]^T$, show that $\mathcal{J}^T\psi(t)\mathcal{J} = \phi(t)$, and hence deduce that $\phi(t)$ is symplectic.

5.3 Riccati algebraic equations

We now consider the situation when the final time t_1 in the performance index (5.2.2) tends to infinity. Since the aim is to make $x(t)$ approach zero as $t \to \infty$, it no longer makes sense to include a terminal expression in the performance index, so we set $M = 0$ in (5.2.2). An important result is due to Kalman [18]:

THEOREM 5.11 If (5.2.1) is controllable† and $M = 0$, then in Theorem 5.8

$$\lim_{t_1 \to \infty} \Pi(t) = \hat{P}(t)$$

exists for all t and is a solution of the Riccati equation (5.2.4).

In the remainder of this section we shall take the matrices in (5.2.1) and (5.2.2) to be real and constant (i.e. independent of t), so that the optimization problem becomes:

Choose u so as to minimize

$$\tfrac{1}{2}\int_0^\infty (x^TQx + u^TRu)\,dt, \tag{5.3.1}$$

subject to

$$\dot{x} = Ax + Bu, \quad x(0) = x_0. \tag{5.3.2}$$

As before, R is symmetric positive definite but we now require Q to be also symmetric positive definite (instead of semidefinite, as in the time-varying case). Notice that there is now no loss of generality in taking

† *See Appendix* 2.

$t_0 = 0$, and the limiting matrix $\hat{P}(t)$ in Theorem 5.11 is constant. We have:

THEOREM 5.12 The solution to the problem (5.3.1) and (5.3.2) is given by

$$u = -R^{-1}B^TPx, \tag{5.3.3}$$

where, if the pair $[A, B]$ is controllable, the constant symmetric matrix P is the unique positive definite solution of the *algebraic matrix Riccati equation*

$$PBR^{-1}B^TP - A^TP - PA - Q = 0, \tag{5.3.4}$$

and the minimum value of (5.3.1) is $\frac{1}{2}x_0^TPx_0$.

Thus, unlike the case when t_1 is finite, the optimal control for a linear time-invariant system with quadratic performance index is also time invariant. Because the solution P of (5.3.4) is the limiting solution of the corresponding differential equation, (5.3.4) is often referred to as the 'steady state' Riccati equation. Notice that we now have to impose the controllability criterion to ensure that the solution to (5.3.4) is positive definite, and hence unique (see Exercise 5.3.1). (Clearly a quadratic algebraic matrix equation will in general have more than one solution.) If Q is only semidefinite and $Q = Q_1^TQ_1$ then $[A, Q_1]$ must be observable (see [21]).

The resulting optimal system when (5.3.3) is applied to (5.3.2) is

$$\dot{x} = (A - BR^{-1}B^TP)x, \tag{5.3.5}$$

and we have the important result:

THEOREM 5.13 The closed loop system (5.3.5), where P is given by Theorem 5.12, is asymptotically stable.

Proof. It is easy to verify, using (5.3.4), that

$$(A - BR^{-1}B^TP)^TP + P(A - BR^{-1}B^TP) = -Q - PBR^{-1}B^TP. \tag{5.3.6}$$

Clearly the matrix on the right in (5.3.6) is positive definite, so that since P is also positive definite the desired result follows at once from Theorems 4.2 and 4.1. In particular, the quadratic form x^TPx is a Liapunov function for (5.3.5).

Kalman [18] has proved a result similar to Theorem 5.13 when the matrices A, B, Q and R are allowed to be time-varying.

Equation (5.2.6) can now be written

$$\left.\begin{aligned} \begin{bmatrix} \dot{x}(t) \\ \dot{p}(t) \end{bmatrix} &= H\begin{bmatrix} x(t) \\ p(t) \end{bmatrix}, \\ H &= \begin{bmatrix} A & -BR^{-1}B^T \\ -Q & -A^T \end{bmatrix} \end{aligned}\right\} \tag{5.3.7}$$

subject to the boundary conditions

$$x(0) = x_0, \, p(\infty) = 0. \tag{5.3.8}$$

MacFarlane [25] has given an explicit expression for the solution of (5.3.7), subject to (5.3.8) (and thereby an expression for the solution of (5.3.4)) in terms of the characteristic roots and vectors of H.

One method of solving (5.3.4) is of course to solve the corresponding differential equation (as in Section 2) with successively larger values of t_1, relying on Theorem 5.11. However, we can in fact give an explicit algebraic solution of algebraic Riccati equations, and we do so first for a general form of the equation obtained from (5.1.1):

$$XEX + DX + XF + G = 0, \tag{5.3.9}$$

where D, E, F, G and X are now all constant $n \times n$ matrices. A matrix corresponding to H in (5.3.7) can be written from (5.1.9) as

$$\alpha = \begin{bmatrix} -F & -E \\ G & D \end{bmatrix}. \tag{5.3.10}$$

The following result demonstrates the nonuniqueness of a solution of (5.3.9).

THEOREM 5.14 [1] Let T be any matrix which transforms α into its Jordan form, so that $T^{-1}\alpha T = \beta$. In partitioned form we write, using an obvious notation,

$$\begin{bmatrix} -F & -E \\ G & D \end{bmatrix} \begin{bmatrix} T_1 & T_2 \\ T_3 & T_4 \end{bmatrix} = \begin{bmatrix} T_1 & T_2 \\ T_3 & T_4 \end{bmatrix} \begin{bmatrix} \beta_1 & \beta_2 \\ 0 & \beta_4 \end{bmatrix}, \tag{5.3.11}$$

where β_1 and β_4 are upper triangular. Then, provided T_1 is nonsingular, the matrix $X = T_3 T_1^{-1}$ is a solution of (5.3.9).

Proof. Multiplying appropriate block rows and columns in (5.3.11) gives

$$FT_1 + ET_3 = -T_1\beta_1 \tag{5.3.12}$$

$$GT_1 + DT_3 = T_3\beta_1. \tag{5.3.13}$$

Assuming T_1 is nonsingular, multiply (5.3.12) by T_1^{-1} on the right and by $T_3 T_1^{-1}$ on the left, and (5.3.13) on the right by T_1^{-1}, and add the resulting equations to get

$$T_3 T_1^{-1} F + T_3 T_1^{-1} E T_3 T_1^{-1} + G + D T_3 T_1^{-1} = 0,$$

which establishes the theorem.

We now return to the Riccati equation (5.3.4) when α in Theorem 5.14 is replaced by the matrix H given in (5.3.7). Under the assumption that the Jordan form of H is diagonal, Potter [31] and O'Donnell [28] have derived the following explicit expression for the unique positive definite solution of (5.3.4).

THEOREM 5.15 Let

$$a_i = \begin{bmatrix} b_i \\ c_i \end{bmatrix},$$

where b_i and c_i are column n-vectors, be a (right) characteristic vector of H in (5.3.7) corresponding to a characteristic root λ_i having negative real part ($i = 1, 2, \ldots, n$). Then

$$P = [c_1, c_2, \ldots, c_n][b_1, b_2, \ldots, b_n]^{-1} \tag{5.3.14}$$

is the positive definite symmetric solution of (5.3.4).

Proof. From Exercises 5.2.5 and 5.2.6, or by realizing that trace $H = 0$, it follows that H has exactly n characteristic roots with negative real parts. It can be shown [25] that in general H must have no purely imaginary roots; this is also necessary for the stability of the closed-loop system (5.3.5) (see Theorem 5.16). Since the Jordan form of H is assumed diagonal, the matrix T in (5.3.11) can therefore be taken as $[a_1, a_2, \ldots, a_n, a_{n+1}, a_{n+2}, \ldots, a_{2n}]$, where $a_{n+1}, \ldots, a_{2n}$ are characteristic vectors corresponding to characteristic roots of H having positive real parts. Thus by Theorem 5.14, the matrix P in (5.3.14) is certainly a solution of (5.3.4), and it remains to show that it is symmetric and positive definite.

First, write

$$P = ([b_1, \ldots, b_n]^{-1})^T P_1 [b_1, \ldots, b_n]^{-1} \tag{5.3.15}$$

where

$$P_1 = [b_1, \ldots, b_n]^T [c_1, \ldots, c_n], \tag{5.3.16}$$

so that P will be symmetric if $P_1 = [p_{ij}]$ is symmetric.

Clearly $p_{ij} = b_i^T c_j$, so that

$$\begin{aligned} p_{ij} - p_{ji} &= c_j^T b_i - b_j^T c_i \qquad i, j = 1, 2, \ldots, n \\ &= [b_j^T \;\; c_j^T] \begin{bmatrix} 0 & -I_n \\ I_n & 0 \end{bmatrix} \begin{bmatrix} b_i \\ c_i \end{bmatrix} \\ &= a_j^T \mathcal{J} a_i, \text{ say.} \end{aligned}$$

Since λ_i and λ_j both have negative real parts, $\lambda_i + \lambda_j \neq 0$, all i, j so we can write

$$\begin{aligned} p_{ij} - p_{ji} &= (\lambda_i a_j^T \mathcal{J} a_i + \lambda_j a_j^T \mathcal{J} a_i)/(\lambda_i + \lambda_j) \\ &= a_j^T (\mathcal{J} H + H^T \mathcal{J}) a_i / (\lambda_i + \lambda_j), \end{aligned} \tag{5.3.17}$$

by using the fact that λ_i, λ_j and a_i, a_j are corresponding characteristic roots and vectors of H. It is easily verified, using (5.3.7) and the definition of $\mathcal{J}$, that the expression on the right in (5.3.17) vanishes (see Exercise 5.2.5), so that P_1 is indeed symmetric.

Finally, we wish to show that P in (5.3.14) is positive definite. Let $U(t) = [\exp(\lambda_1 t) a_1, \ldots, \exp(\lambda_n t) a_n]$ so that

$$\dot{U}(t) = H U(t). \tag{5.3.18}$$

If we define the $2n \times 2n$ matrix

$$K = \begin{bmatrix} 0 & I_n \\ 0 & 0 \end{bmatrix},$$

then from (5.3.16) we have

$$P_1 = U^T(0)KU(0),$$

and since all the λ_i have negative real parts, clearly $U(t) \to 0$ as $t \to \infty$, so we can write

$$P_1 = -\int_0^\infty \frac{d}{dt}[U^T(t)KU(t)]\, dt$$

$$= \int_0^\infty U^T(t)[-H^TK - KH]U(t)\, dt \tag{5.3.19}$$

by using (5.3.18). However, from (5.3.7) we obtain

$$H^TK + KH = -\begin{bmatrix} Q & 0 \\ 0 & BR^{-1}B^T \end{bmatrix},$$

which is negative definite, showing that the integrand in (5.3.19) is positive definite for all $t > 0$, so that P_1 (and hence P) must be positive definite.

Theorem 5.12 and Exercise 5.3.1 ensure that the positive definite solution of equation (5.3.4) exists and is unique.

Theorem 5.15 enables us to demonstrate the relationship between the characteristic roots of the closed-loop system (5.3.5) and those of H:

THEOREM 5.16 [28] Under the same assumptions as in Theorem 5.15, the characteristic roots of the closed-loop system (5.3.5) are those characteristic roots of H having negative real parts.

Proof. Since $a_1, \ldots, a_{2n}$ are the characteristic vectors of H and its Jordan form is assumed diagonal, we can write

$$\begin{bmatrix} A & -BR^{-1}B^T \\ -Q & -A^T \end{bmatrix} \begin{bmatrix} b_1 & \cdots & b_n & \cdots & b_{2n} \\ c_1 & \cdots & c_n & \cdots & c_{2n} \end{bmatrix}$$

$$= \begin{bmatrix} b_1 & \cdots & b_n & \cdots & b_{2n} \\ c_1 & \cdots & c_n & \cdots & c_{2n} \end{bmatrix} \begin{bmatrix} \lambda_1 & & & & \\ & \ddots & & & \\ & & \lambda_n & & \\ & & & \ddots & \\ & & & & \lambda_{2n} \end{bmatrix}. \tag{5.3.20}$$

By regarding the matrices in (5.3.20) as being partitioned into $n \times n$ blocks, and by equating the (21) elements in the partitioned multiplication we obtain

$$A[b_1, \ldots, b_n] - BR^{-1}B^T[c_1, \ldots, c_n] = [b_1, \ldots, b_n] \operatorname{diag}[\lambda_1, \ldots, \lambda_n], \tag{5.3.21}$$

or

$$A - BR^{-1}B^T[c_1, \ldots, c_n]\,[b_1, \ldots, b_n]^{-1} = [b_1, \ldots, b_n]\ \mathrm{diag}\,[\lambda_1, \ldots, \lambda_n]\,[b_1, \ldots, b_n]^{-1}$$

which shows that $A - BR^{-1}B^TP$ is similar to diag $[\lambda_1, \ldots, \lambda_n]$, and these λ_i are just the n characteristic roots of H having negative real parts.

Although we already knew, from Theorem 5.13, that the closed-loop system is asymptotically stable, it is important to be able to identify its characteristic roots. The characteristic vectors of the closed-loop system are also easily obtained (see Exercise 5.3.5). O'Donnell [28] claims that Theorem 5.15 is effective when applied numerically and Fath [12], who uses a slight modification, also reports satisfactory computational experience.

Some results due to Roth [35] are worth giving here. First notice that if E in (5.3.9) is nonsingular we can write that equation as

$$Y^2 + DY + YE^{-1}FE + GE = 0, \tag{5.3.22}$$

where $Y = XE$. If we then put $Y = Y_1 - D$, (5.3.22) becomes

$$Y_1^2 + Y_1(E^{-1}FE - D) + GE - DE^{-1}FE = 0,$$

and this is an example of a *unilateral* matrix equation (see [15], p. 227, [24], p. 95, and [17]). For the Riccati equation (5.3.4), Freested *et al.* [13] have applied Roth's work to obtain the following:

Let $\Delta(\lambda) = \det(\lambda I_{2n} - H)$, where H is given by (5.3.7). As we have seen in the proof of Theorem 5.15, we can factorize so that $\Delta(\lambda) = (-1)^n\Delta_1(\lambda)\Delta_1(-\lambda)$, with $\Delta_1(\lambda)$ a Hurwitz polynomial. Write also

$$\Delta_1(H) = \begin{bmatrix} H_1 & H_2 \\ H_3 & H_4 \end{bmatrix}.$$

THEOREM 5.17 The positive definite solution P of (5.3.4) is given by the solution of

$$\begin{bmatrix} H_1 & H_2 \\ H_3 & H_4 \end{bmatrix}\begin{bmatrix} I_n \\ P \end{bmatrix} = \begin{bmatrix} 0 \\ 0 \end{bmatrix}. \tag{5.3.23}$$

Proof. If we define

$$P_2 = \begin{bmatrix} I_n & 0 \\ P & I_n \end{bmatrix},$$

then

$$\begin{aligned} P_2^{-1}HP_2 &= \begin{bmatrix} I_n & 0 \\ -P & I_n \end{bmatrix}\begin{bmatrix} A & -BR^{-1}B^T \\ -Q & -A^T \end{bmatrix}\begin{bmatrix} I_n & 0 \\ P & I_n \end{bmatrix} \\ &= \begin{bmatrix} A_1 & -BR^{-1}B^T \\ Z & -A_1^T \end{bmatrix} \end{aligned} \tag{5.3.24}$$

where

$$A_1 = A - BR^{-1}B^TP, \qquad Z = PBR^{-1}B^TP - A^TP - PA - Q.$$

Also, (5.3.23) gives

$$\left.\begin{aligned} H_1 + H_2P = 0 \\ H_3 + H_4P = 0 \end{aligned}\right\}. \tag{5.3.25}$$

By using (5.3.25) we get

$$\begin{bmatrix} I_n & 0 \\ -P & I_n \end{bmatrix}\begin{bmatrix} H_1 & H_2 \\ H_3 & H_4 \end{bmatrix}\begin{bmatrix} I_n & 0 \\ P & I_n \end{bmatrix} = \begin{bmatrix} 0 & X \\ 0 & X \end{bmatrix}$$

(where X denotes a nonzero matrix) or

$$P_2^{-1}\Delta_1(H)P_2 = \begin{bmatrix} 0 & X \\ 0 & X \end{bmatrix},$$

which implies

$$\Delta_1(P_2^{-1}HP_2) = \begin{bmatrix} 0 & X \\ 0 & X \end{bmatrix}. \tag{5.3.26}$$

From (5.3.24) and (5.3.26) we conclude that $Z = 0$, which shows that P is a solution of (5.3.4). From (5.3.24) we can also deduce that

$$\begin{aligned} \det(\lambda I_{2n} - P_2^{-1}HP_2) &= \Delta(H) \\ &= \det(\lambda I_n - A_1)\det(\lambda I_n + A_1^T), \end{aligned}$$

so that $\det(\lambda I_n - A_1) = \Delta_1(\lambda)$. Hence A_1 will be a stability matrix and P will be positive definite (Theorem 5.16).

An entirely different procedure for finding the solution P of (5.3.4) has been given by Kleinman [22]:

THEOREM 5.18 Let P_k $(k = 0, 1, 2, \ldots)$ be the unique positive definite solution of the linear matrix equation

$$A_k^TP_k + P_kA_k = -Q - P_{k-1}BR^{-1}B^TP_{k-1} \tag{5.3.27}$$

where $A_k = A - BR^{-1}B^TP_{k-1}$, $k = 1, 2, \ldots$; $A_0 = A - BL_0$, and L_0 is any matrix such that $A - BL_0$ is a stability matrix. Then $\lim_{k\to\infty} P_k = P$, and† $P \leqslant \cdots \leqslant V_{k+2} \leqslant V_{k+1} \leqslant V_k \leqslant \cdots$

Notice that since $[A, B]$ is assumed controllable, Theorem 4.8 ensures the existence of a suitable matrix L_0. Kleinman shows that P_k is quadratically convergent, and in fact the iterative scheme of his theorem can be obtained by applying Newton's method in function space. The matrix equation (5.3.27) is of the type discussed in Section 4.1, and may be solved without too much difficulty (a good method is one based on using an infinite matrix series [7]). One drawback to the use of Theorem 5.18 is the difficulty of finding a matrix L_0 such that P_0 is a sufficiently good approximation to P for convergence to be rapid. Man [26] has proposed a method using function minimization which is still quadratically convergent but does not depend so critically on P_0.

† *Using the same notation as in the footnote on page* 118.

As in the case of the Riccati differential equation, we do not propose to comment on comparative efficiencies of methods for solving the steady state equation, but refer the reader to [11], [13] and [20] for reported experience.

We now briefly look again at relationships discovered by Anderson [2, 3] between Riccati equations and the spectral factorization problem (see Section 3.4). First notice that in Theorem 5.10, if

$$Z(\lambda) = \tfrac{1}{2}R + S^T(\lambda I_n - A)^{-1}B \tag{5.3.28}$$

is positive real, then in equation (5.2.16)

$$\lim_{t_1 \to \infty} \rho(t) = \hat{\rho} \tag{5.3.29}$$

exists, is independent of t and is negative definite provided $[A, S]$ is observable (compare with Theorem 5.11). Using the notation of equation (3.4.2), we wish to find a matrix $W_1(\lambda)$ such that

$$Z(\lambda) + Z^T(-\lambda) = W_1^T(-\lambda)W(\lambda),$$

where $Z(\lambda)$ is the positive-real matrix in (5.3.28). Anderson gives a solution to this factorization problem in terms of the limiting solution $\hat{\rho}$ in (5.3.29). This contrasts with Theorem 3.28 which involves finding the solution of the algebraic equation (see Exercise 3.4.9):

$$\begin{aligned} P_1BY_0^{-1}B^TP_1 &+ (A - BY_0^{-1}C)^TP_1 \\ &+ P_1(A - BY_0^{-1}C) + C^TY_0^{-1}C = 0, \end{aligned} \tag{5.3.30}$$

subject to the condition that $A + BY_0^{-1}(B^TP_1 - C)$ be a stability matrix. It can be shown [2] that the desired solution of (5.3.30) can be obtained from Theorem 5.14 provided β_1 in equation (5.3.11) is associated with characteristic roots of α in (5.3.10) (with appropriate D, E, F, G) having *positive* real parts (see Exercise 5.3.7). This difference from Theorem 5.15 is due to the fact that the constant term $C^TY_0^{-1}C$ in (5.3.30) has opposite sign to that in the usual Riccati algebraic equation (5.3.4).

Finally we quote a result of Reid [34] for the following special case of (5.3.4):

$$PBR^{-1}B^TP - A^TP - PA = 0 \tag{5.3.31}$$

THEOREM 5.19 If $[A, BR^{-1}B^T]$ is controllable, then there exist symmetric matrices $P^{(1)}$ and $P^{(2)}$ satisfying (5.3.31) which are extreme solutions in the sense that if P is any symmetric matrix satisfying (5.3.31) then† $P^{(1)} \leqslant P \leqslant P^{(2)}$.

EXERCISES

5.3.1 If P_1 and P_2 are two positive definite solutions of equation (5.3.4), show that

† *Again using the notation of the footnote on page* 118.

$$(P_1-P_2)(A-BR^{-1}B^TP_1)+(A^T-P_2BR^{-1}B^T)(P_1-P_2)=0.$$

Hence deduce, using Theorems 5.13 and 1.9, that a positive definite solution of (5.3.4), if it exists, is unique.

5.3.2 Set $B = 0$ in (5.3.4) and show that the characteristic roots of the corresponding matrix H_0 in (5.3.7) are just $\pm\mu_i$, where μ_i $(i = 1, 2, \ldots, n)$ are the roots of A (assumed distinct). Hence, using Theorem 5.15, show that if A is a stability matrix the solution of the Liapunov matrix equation $A^TP + PA = -Q$ is $P = [c_1, \ldots, c_n][b_1, \ldots, b_n]^{-1}$ where b_i is a characteristic vector of A corresponding to μ_i and $c_i = -(A^T + \mu_i I_n)^{-1}Qb_i$.

5.3.3 Show that the matrix P in the previous exercise can be written $P = -(B^T)^{-1}P_1B^{-1}$ where $B = [b_1, \ldots, b_n]$ and $P_1 = [p_{ij}]$ with $p_{ij} = (b_i{}^TQb_j)/(\mu_i + \mu_j) = p_{ji}$.

5.3.4 Show that the solution in the previous exercise of $A^TP + PA = -Q$ can be obtained by diagonalizing A.

5.3.5 Using the method of Theorem 5.16, show that the characteristic vectors of the closed-loop system matrix in equation (5.3.5) are $[b_1, \ldots, b_n]e_i$, $i = 1, 2, \ldots, n$, where e_i is the ith column of I_n.

5.3.6 Show that the $n \times n$ matrix X is a solution of the *unilateral* matrix equation

$$X^pA_0 + X^{p-1}A_1 + \cdots + A_p = 0$$

(so-called since multiplication of each power of X is on the same side) if and only if $\lambda I_n - X$ is a left factor of

$$A(\lambda) = A_0\lambda^p + A_1\lambda^{p-1} + \cdots + A_p$$

(see Exercise 1.1.1).

5.3.7 Apply the method of Theorem 5.14 to equation (5.3.30) to show that $A + BY_0^{-1}(B^TP_1 - C)$ is similar to $-\beta_1$.

REFERENCES

1. ANDERSON, B. D. O., 'Solution of quadratic matrix equations', *Electronics Letters*, **2**, 371–372 (1966).
2. ANDERSON, B. D. O., 'An algebraic solution to the spectral factorization problem', *IEEE Trans. Aut. Control*, **AC-12**, 410–414 (1967).
3. ANDERSON, B. D. O., 'Quadratic minimization, positive real matrices and spectral factorization', *Tech. Rept. EE-6812*, University of Newcastle, New South Wales (1968).
4. ASSEO, S. J., 'Optimal control of a servo derived from nonquadratic performance criteria', *IEEE Trans. Aut. Control*, **AC-14**, 404–407 (1969).
5. ATHANS, M. and FALB, P. L., *Optimal control*, McGraw-Hill, New York (1966).

6. BARNETT, S. and STOREY, C., 'Insensitivity of optimal linear control systems to persistent changes in parameters', *Int. J. Control*, **4**, 179–184 (1966).
7. BARNETT, S. and STOREY, C., *Matrix methods in stability theory*, Nelson, London (1970).
8. BASS, R. W. and WEBBER, R. F., 'Optimal nonlinear feedback control derived from quartic and higher-order performance criteria', *IEEE Trans. Aut. Control*, **AC-11**, 448–454 (1966).
9. BELLMAN, R., *Introduction to matrix analysis*, 2nd Edn., McGraw-Hill, New York (1970).
10. BELLMAN, R. 'Upper and lower bounds for the solutions of the matrix Riccati equation', *J. Math. Anal. Appl.* **17**, 373–379 (1967).
11. BLACKBURN, T. R. and BIDWELL, J. C., 'Some numerical aspects of control engineering computations', *Preprints Joint Automatic Control Conference*, Ann Arbor, Michigan, 203–207 (1968).
12. FATH, A. F., 'Computational aspects of the linear optimal regulator problem', *IEEE Trans. Aut. Control*, **AC-14**, 547–550 (1969).
13. FREESTED, W. C., WEBBER, R. F. and BASS, R. W., 'The "GASP" computer program—an integrated tool for optimal control and filter design', *Preprints Joint Automatic Control Conference*, Ann Arbor, Michigan, 198–202 (1968).
14. FRIEDLAND, B., 'On solutions of the Riccati equation in optimization problems', *IEEE Trans. Aut. Control*, **AC-12**, 303–304 (1967).
15. GANTMACHER, F. R., *Theory of matrices*, Vol. I, Chelsea Publishing Company, New York (1960).
16. INCE, E. L., *Integration of ordinary differential equations*, 7th Edn., Oliver and Boyd, Edinburgh (1963).
17. INGRAHAM, M. H., 'Rational methods in matrix equations', *Bull. Amer. Math. Soc.* **47**, 61–70 (1941).
18. KALMAN, R. E., 'Contributions to the theory of optimal control', *Bol. Soc. Mat. Mex.* **5**, 102–119 (1960).
19. KALMAN, R. E. and BUCY, R. S., 'New results in linear filtering and prediction theory', *Trans. ASME, J. Basic Eng.* **83D**, 95–108 (1961).
20. KALMAN, R. E. and ENGLAR, T. S., 'A user's manual for the automatic synthesis program', *NASA Report CR-475*, Washington (1966).
21. KLEINMAN, D. L., 'On the linear regulator problem and the matrix Riccati equation', *Report ESL-R-271*, Massachusetts Institute of Technology (1966).
22. KLEINMAN, D. L., 'On an iterative technique for Riccati equation computations', *IEEE Trans. Aut. Control*, **AC-13**, 114–115 (1968).

23. LEVIN, J. J., 'On the matrix Riccati equation', *Proc. Amer. Math. Soc.* **10,** 519–524 (1959).
24. MACDUFFEE, C. C., *The theory of matrices,* Chelsea Publishing Company, New York (1956).
25. MACFARLANE, A. G. J., 'An eigenvector solution of the optimal linear regulator problem', *J. Electron. Contr.* **14,** 643–654 (1963).
26. MAN, F. T., 'The Davidson method of solution of the algebraic matrix Riccati equation', *Int. J. Control,* **10,** 713–719 (1969).
27. MAN, F. T. and SMITH, H. W., 'Design of linear regulators optimal for time-multiplied performance indices', *IEEE Trans. Aut. Control,* **AC-14,** 527–529 (1969).
28. O'DONNELL, J. J., 'Asymptotic solution of the matrix Riccati equation of optimal control', *Proc. 4th Annual Allerton Conference on circuit and system theory,* University of Illinois, Urbana, 577–586 (1966).
29. PIAGGIO, H. T. H., *An elementary treatise on differential equations and their applications,* Bell, London (1954).
30. PORTER, W. A., 'On the matrix Riccati equation', *IEEE Trans. Aut. Control,* **AC-12,** 746–749 (1967).
31. POTTER, J. E., 'Matrix quadratic solutions', *SIAM J. Appl. Math.* **14,** 496–501 (1966).
32. REID, W. T., 'A matrix differential equation of Riccati type', *Amer. J. Math.* **68,** 237–246 (1946).
33. REID, W. T., 'Riccati matrix differential equations and non-oscillation criteria for associated linear differential systems', *Pacific J. Math.* **13,** 665–685 (1963).
34. REID, W. T., 'A matrix equation related to a non-oscillation criterion and Liapunov stability', *Quart. Applied Math.* **23,** 83–87 (1965).
35. ROTH, W. E., 'On the matrix equation $X^2 + AX + XB + C = 0$', *Proc. Amer. Math. Soc.* **1,** 586–589 (1950).
36. RUGH, W. J. and MURPHY, G. J., 'A new approach to the solution of linear optimal control problems', *Trans. ASME, J. Basic Eng.* **91D,** 149–154 (1969).
37. VAUGHAN, D. R., 'A negative exponential solution for the matrix Riccati equation', *IEEE Trans. Aut. Control,* **AC-14,** 72–75 (1969).
38. WHYBURN, W. M., 'Matrix differential equations', *Amer. J. Math.* **56,** 587–592 (1934).
39. WONHAM, W. M., 'On pole assignment in multi-input controllable linear systems', *IEEE Trans. Aut. Control,* **AC-12,** 660–665 (1967).

6

Generalized Inverses

6.1 The Moore–Penrose inverse

The idea of generalizing the concept of inverse to cover singular and rectangular matrices has been found very useful in statistics and in a number of areas of applied mathematics, but rather than go into details of specific applications we refer the reader to two survey papers [2, 26] which list many references.

It is helpful to begin by considering the system of linear equations

$$Ax = b, \tag{6.1.1}$$

where the matrix A, and column vectors x and b may have complex elements. If A is square and nonsingular then the solution of (6.1.1) is of course simply $x = A^{-1}b$. However if A is singular or rectangular the equations will not in general have a unique solution and indeed may not even be consistent. In these cases let us specify that x is to be chosen so as to give a least-squares solution of (6.1.1). That is, if A is $m \times n$, choose x so that the sum of squares

$$\begin{aligned}\|Ax - b\|^2 &= (Ax - b)^*(Ax - b)\\ &= x^*A^*Ax - x^*A^*b - b^*Ax + b^*b\end{aligned} \tag{6.1.2}$$

is a minimum. Differentiating (6.1.2) with respect to x and equating to zero gives

$$A^*Ax = A^*b,$$

so that if A^*A is nonsingular (i.e. if A has rank n—see Exercise 6.1.1) we have the unique solution $x = Bb$ to the stated problem, where

$$B = (A^*A)^{-1}A^*. \tag{6.1.3}$$

The $n \times m$ matrix B is a left inverse of A, since $BA = I_n$, and if A is nonsingular B reduces to A^{-1}.

Next suppose that the $m \times n$ matrix A has rank $r > 0$. It is always possible to find matrices such that

$$A = CD \tag{6.1.4}$$

where C is $m \times r$, D is $r \times n$ and both have rank r. Moore [29] essentially defined a *generalized inverse*† (g.i.) of A by means of

$$A^+ = D^*(DD^*)^{-1}(C^*C)^{-1}C^* \tag{6.1.5}$$

(we follow here [17]). When $r = n$, D reduces to I_n and A^+ is then identical to B in (6.1.3). Similarly, when $r = m$, A^+ reduces to a right inverse of A (Exercise 6.1.2). Quite independently Penrose [32] rediscovered this generalized inverse in the following form:

THEOREM 6.1 The $n \times m$ matrix X satisfying the four equations

$$AXA = A \tag{6.1.6}$$

$$XAX = X \tag{6.1.7}$$

$$(AX)^* = AX \tag{6.1.8}$$

$$(XA)^* = XA \tag{6.1.9}$$

exists and is unique.

Proof. It is very easy to verify that the matrix A^+ in (6.1.5) satisfies (6.1.6)–(6.1.9).

To show uniqueness, let X and Y be two matrices satisfying (6.1.6)–(6.1.9). Then

$$\begin{aligned} X = XAX &= A^*X^*X = A^*Y^*A^*X^*X = A^*Y^*XAX \\ &= A^*Y^*X = YAX = YX^*A^* = YAYX^*A^* \\ &= YY^*A^*X^*A^* = YY^*A^* = YAY = Y. \end{aligned}$$

A^+ is often called the *Moore–Penrose* generalized inverse of A and Rado [35] has shown the equivalence of the definitions of these two authors; yet another independent derivation was given by Bjerhammar [5, 6]. If A is a zero matrix we define A^+ to be an $n \times m$ zero matrix, so in particular for a scalar a, $a^+ = 1/a$, $a \neq 0$ and $a^+ = 0$ if $a = 0$. Clearly $A^+ = A^{-1}$ if A is nonsingular, but notice that A^+ may not be a continuous function of A, as the following simple example illustrates:

$$A = \begin{bmatrix} a & 1 \\ 1 & 1 \end{bmatrix}, \; A^{-1} = \frac{1}{a-1}\begin{bmatrix} 1 & -1 \\ -1 & a \end{bmatrix}, \qquad a \neq 1,$$

$$A^+ = \frac{1}{4}\begin{bmatrix} 1 & 1 \\ 1 & 1 \end{bmatrix}, \qquad a = 1.$$

However, if the rank of A does not alter then A^+ does vary continuously with A [32].

Using the defining equations (6.1.6)–(6.1.9) it is easy to show that A^+ has the following properties.

THEOREM 6.2 [32]

(i) If A^{-1} exists, $A^+ = A^{-1}$.

† *Some authors use the term 'pseudo-inverse'.*

(ii) $(A^+)^+ = A$.
(iii) $(A^*)^+ = (A^+)^*$.
(iv) $(kA)^+ = k^+A^+$ for any scalar k.
(v) $(A^*A)^+ = A^+(A^+)^*$
(vi) If U and V are unitary, $(UAV)^+ = V^*A^+U^*$.
(vii) If $A = \sum_i A_i$ and $A_iA_j^* = 0 = A_i^*A_j$, all $i \neq j$, then
$A^+ = \sum_i A_i^+$.
(viii) If A is normal, $A^+A = AA^+$ and $(A^p)^+ = (A^+)^p$ for positive integers p.
(ix) A, A^*A, A^+ and A^+A all have rank equal to trace (A^+A).

Notice that (v) does not hold in general; i.e., $(AB)^+ \neq B^+A^+$ for arbitrary matrices A and B conformable for multiplication, although equality does hold if, for example, A has full column rank and B has full row rank (see Exercise 6.1.11). For the general case Greville [18] has proved:

THEOREM 6.3 Each of the following is a necessary and sufficient condition for $(AB)^+ = B^+A^+$.

(i) $A^+ABB^*A^* = BB^*A^*$ and $BB^+A^*AB = A^*AB$
(ii) A^+ABB^* and A^*ABB^+ are Hermitian
(iii) $A^+ABB^*A^*ABB^+ = BB^*A^*A$.

It is worth pointing out that an alternative proof of the existence of the g.i. of any square matrix A can be obtained from the result that A can be written in the form VEW where V and W are unitary and E diagonal. The matrix E^+ is defined in an obvious fashion (see Exercise 6.1.3) and it is then sufficient to verify that $W^*E^+V^* = A^+$ [32]. A rectangular matrix can be dealt with by bordering it with zeros to make it square.

We now return to the solution of linear equations.

THEOREM 6.4 [32] A necessary and sufficient condition for the equation

$$AXC = B \qquad (6.1.10)$$

to have a solution is

$$AA^+BC^+C = B \qquad (6.1.11)$$

in which case the general solution is

$$X = A^+BC^+ + Y - A^+AYCC^+, \qquad (6.1.12)$$

where Y is an arbitrary matrix having the same dimensions as X.

Proof. If X satisfies (6.1.10), then

$$B = AXC = AA^{+}AXCC^{+}C = AA^{+}BC^{+}C.$$

Conversely, if (6.1.11) holds then $X = A^{+}BC^{+}$ is a particular solution of (6.1.10).

That (6.1.12) is the general solution of (6.1.10) follows since $A(Y - A^{+}AYCC^{+})C = 0$.

Notice that the only property used in this proof is (6.1.6); we shall refer to this point again in the next section.

We next consider the case when C in (6.1.10) is a unit matrix, and seek the *best approximate solution* X_0 of

$$AX = B, \tag{6.1.13}$$

where by this we mean [33] that for all X, either

$$\|AX - B\| > \|AX_0 - B\| \tag{6.1.14}$$

or

$$\|AX - B\| = \|AX_0 - B\| \text{ and } \|X\| \geqslant \|X_0\|. \tag{6.1.15}$$

Thus X_0 is the matrix having least norm which minimizes the sum of squares $\|AX - B\|^2$. Penrose [33] has proved the following generalization of our earlier simple result:

THEOREM 6.5 The best approximate solution of (6.1.13) is $A^{+}B$.

Proof. Using $A^*AA^{+} = A^*$, it is easy to verify that

$$[AP + (I_m - AA^{+})Q]^*[AP + (I_m - AA^{+})Q] = (AP)^*AP + [(I_m - AA^{+})Q]^*[(I_m - AA^{+})Q]$$

for suitably dimensioned matrices P and Q. That is,

$$\|AP + (I_m - AA^{+})Q\|^2 = \|AP\|^2 + \|(I_m - AA^{+})Q\|^2. \tag{6.1.16}$$

In particular,

$$\begin{aligned}\|AX - B\|^2 &= \|A(X - A^{+}B) + (I_m - AA^{+})(-B)\|^2 \\ &= \|A(X - A^{+}B)\|^2 + \|(I_m - AA^{+})(-B)\|^2 \\ &\geqslant \|AA^{+}B - B\|^2\end{aligned} \tag{6.1.17}$$

with equality only when the first term vanishes, i.e. when $AX = AA^{+}B$. Replacing A by A^{+} in (6.1.16) produces

$$\|A^{+}B + (I_m - A^{+}A)X\|^2 = \|A^{+}B\|^2 + \|(I_m - A^{+}A)X\|^2. \tag{6.1.18}$$

Thus if $AX = AA^{+}B$, (6.1.18) gives

$$\|X\|^2 = \|A^{+}B\|^2 + \|X - A^{+}B\|^2. \tag{6.1.19}$$

The result now follows, since from (6.1.17) and (6.1.19) we see that $X_0 = A^{+}B$ satisfies the conditions of (6.1.14) and (6.1.15).

We now make some comments based on the two previous theorems.

(i) The general solution of (6.1.1) is

$$x = A^{+}b + (I_n - A^{+}A)y, \tag{6.1.20}$$

where y is an arbitrary column n-vector, provided (6.1.1) has a solution (i.e. provided $AA^{+}b = b$).

(ii) The best approximate solution of (6.1.1) is $x = A^{+}b$, which confirms and generalizes our elementary approach at the beginning of this section.

(iii) The best approximate solution of $AX = I_n$ is $X = A^{+}$, which suggests an alternative definition of the Moore–Penrose g.i.

(iv) A necessary and sufficient condition for the pair of equations $AX = B$, $XD = E$ to have a solution X is that each equation should individually have a solution and that $AE = BD$. Necessity is obvious, and sufficiency follows by taking

$$X = A^{+}B + ED^{+} - A^{+}AED^{+} \text{ [32].}$$

(v) Using a similar argument, Theorem 6.5 can be extended to show that the best approximate solution of (6.1.10) is $A^{+}BC^{+}$.

We have already seen that if A has full rank (so that either A^*A or AA^* is nonsingular) then a simple expression for A^{+} can be written down at once. If however A has rank $r < \min(m, n)$ then we can use (6.1.5) or proceed by partitioning as follows [33]. After suitable equivalence transformations A can be written as

$$\begin{bmatrix} A_1 & A_2 \\ A_3 & A_4 \end{bmatrix} \tag{6.1.21}$$

where A_1 is $r \times r$ and nonsingular. There exist nonsingular matrices X, Y such that

$$\begin{bmatrix} A_1 & A_2 \\ A_3 & A_4 \end{bmatrix} = \begin{bmatrix} X_1 & X_2 \\ X_3 & X_4 \end{bmatrix} \begin{bmatrix} I_r & 0 \\ 0 & 0 \end{bmatrix} \begin{bmatrix} Y_1 & Y_2 \\ Y_3 & Y_4 \end{bmatrix} \tag{6.1.22}$$

where X and Y have been partitioned in an obvious fashion.

Carrying out block multiplication in (6.1.22) gives

$$A_1 = X_1Y_1,\ A_2 = X_1Y_2,\ A_3 = X_3Y_1,\ A_4 = X_3Y_2$$

so that $A_4 = A_3Y_1^{-1}X_1^{-1}A_2 = A_3A_1^{-1}A_2$ (X_1 and Y_1 are nonsingular since A_1 is nonsingular). It is then straightforward to verify using Theorem 6.1 that

$$\begin{bmatrix} A_1 & A_2 \\ A_3 & A_3A_1^{-1}A_2 \end{bmatrix}^{+} = \begin{bmatrix} A_1^*KA_1^* & A_1^*KA_3^* \\ A_2^*KA_1^* & A_2^*KA_3^* \end{bmatrix},$$

where $K = [A_1A_1^* + A_2A_2^*]^{-1}A_1[A_1^*A_1 + A_3^*A_3]^{-1}$. The matrices $A_1A_1^* + A_2A_2^*$ and $A_1^*A_1 + A_3^*A_3$ are in fact positive definite since A_1 is nonsingular. In the same paper Penrose suggests a finite iterative procedure for finding A^{+}.

Greville [17] also describes a recursive method for finding A^+ which involves partitioning. Let a_i denote the ith column of A and let $A_k = [a_1, a_2, \ldots, a_k]$. Then

$$A_k^+ = \begin{bmatrix} A_{k-1}^+ - d_k b_k \\ b_k \end{bmatrix}$$

where $d_k = A_{k-1}^+ a_k$, $c_k = a_k - A_{k-1} d_k$ and

$$b_k = c_k^+ + (1 - c_k^+ c_k)(1 + d_k^* d_k)^{-1} d_k^* A_{k-1}^+.$$

Notice that the first term in the expression for b_k vanishes if $c_k = 0$, and the second if $c_k \neq 0$. Thus $A_1^+, A_2^+, \ldots, A_n^+ = A^+$ can be found successively. Cline [8] has extended this by giving an expression for the g.i. of $[A_k, B_k]$ where $B_k = [a_{k+1}, \ldots, a_l]$. (See also Theorem 6.11.)

Yet another iterative procedure has been given by Ben-Israel [1]:

THEOREM 6.6 The sequence defined by

$$X_0 = pA^*, X_{k+1} = X_k(2I_m - AX_k) \quad (k = 0, 1, 2, \ldots),$$

converges to A^+ as $k \to \infty$ provided the real scalar p satisfies $0 < p < 2/\mu$, where μ is the largest characteristic root of AA^*.

Alternatively, μ can be replaced (applying Gershgorin's Theorem) by

$$\max_i \sum_{j=1}^{m} |b_{ij}|, \text{ where } AA^* = [b_{ij}].$$

The Cayley–Hamilton Theorem has also been used to obtain an expression for A^+, thus generalizing the result that if A is nonsingular and has the characteristic equation

$$\lambda^n + a_1\lambda^{n-1} + \cdots + a_n = 0$$

then of course

$$A^{-1} = -(A^{n-1} + a_1 A^{n-2} + \cdots + a_{n-1} I_n)/a_n.$$

If A is any $m \times n$ matrix, let the characteristic equation of AA^* be

$$\lambda^m + q_1\lambda^{m-1} + \cdots + q_{m-1}\lambda + q_m = 0,$$

and let k be the largest nonzero integer such that $q_k \neq 0$. Then by applying the Cayley–Hamilton Theorem to AA^* and using the properties of A^+, Decell [11] has proved:

THEOREM 6.7

$$A^+ = -A^*[(AA^*)^{k-1} + q_1(AA^*)^{k-2} + \cdots + q_{k-1} I_m]/q_k.$$

It is then easy to apply Leverrier's algorithm (see Exercise 3.2.3) to AA^* to obtain

$$A^+ = -A^* B_{k-1}/q_k$$

where $B_1 = AA^* + q_1 I_m$, $B_r = AA^* B_{r-1} + q_r I_m$, $r = 2, 3, \ldots, k$.

We have now given a brief account of the basic and more commonly used properties of the Moore–Penrose g.i. For further details and results the references previously mentioned should be consulted (see also [2] for a thorough treatment including g.i. of linear operators). In addition, material on numerical computation of the g.i. can be found in [4] and [27].

EXERCISES

6.1.1 If A is an $m \times n$ matrix show that rank $(A^*A) =$ rank A by considering related systems of homogeneous linear equations.

6.1.2 Obtain the right inverse of A corresponding to B in equation (6.1.3), when A has rank m.

6.1.3 Find A^+ if $A = \text{diag}\,[a_1, a_2, \ldots, a_n]$. Hence obtain an expression for the Moore–Penrose g.i. of a Hermitian matrix.

6.1.4 Verify (i)–(iv) of Theorem 6.2 by using the defining equations in Theorem 6.1.

6.1.5 Show that if $\mathcal{J}$ is an *idempotent* matrix (i.e. $\mathcal{J}^2 = \mathcal{J}$) then rank $\mathcal{J} =$ trace $\mathcal{J}$. Apply this to A^+A and hence using (6.1.5) verify (ix) of Theorem 6.2.

6.1.6 Prove that

$$A^*AA^+ = A^* = A^+AA^*$$

and

$$A^*(A^+)^*A^+ = A^+ = A^+(A^+)^*A^*.$$

6.1.7 Show that $I_n - A^+A$ and $I_m - AA^+$ are Hermitian and idempotent.

6.1.8 Prove that if H is Hermitian and idempotent then $H^+ = H$.

6.1.9 Verify that $A^+ = (A^*A)^+A^*$ by showing that the defining equations of Theorem 6.1 are satisfied (the results in Exercise 6.1.6 are needed).

This shows that the g.i. of an arbitrary rectangular matrix can be expressed in terms of the g.i. of a positive semidefinite Hermitian matrix.

6.1.10 [33] If D satisfies the equation $A^*A = D(A^*A)^2$ show by multiplying on the right by $A^+(A^+)^*A^+$ that $A^+ = DA^*$ (again use Exercise 6.1.6).

6.1.11 If A is $m \times r$, B is $r \times n$ and both matrices have rank r use equation (6.1.5) to show that $(AB)^+ = B^+A^+$.

6.1.12 [11] Show that if $k = 0$ in Theorem 6.7 then $A^+ = 0$.

6.2 Other definitions and properties

The Moore–Penrose g.i. of an $m \times n$ matrix A is not the only one which is useful and in this section we look at some other forms of inverse.

Most of these are based on the four equations in Theorem 6.1, which we repeat here for convenience.

$$AXA = A \tag{6.2.1}$$

$$XAX = X \tag{6.2.2}$$

$$(AX)^* = AX \tag{6.2.3}$$

$$(XA)^* = XA. \tag{6.2.4}$$

We shall denote by $A(i, j, \ldots, l)$ the set of $n \times m$ matrices X satisfying equations (6.2.i), (6.2.j), ..., (6.2.l) ($i, j, \ldots, l = 1, 2, 3, 4$). If $X \in A(i, j, \ldots, l)$ we shall call X *an* $(i, j, \ldots, l)$ g.i. of A and write $A^{(i, j, \ldots, l)}$ for a member of $A(i, j, \ldots, l)$. By definition the (1, 2, 3, 4) g.i. of A is the unique Moore–Penrose g.i. A^+ and since $A(1, 2, 3, 4)$ is nonempty so are all the other sets. Terminology is not entirely standardized: for example, Rohde [37] calls a (1) g.i. simply a generalized inverse; a (1, 2) g.i. a *semi-inverse* or *reflexive* g.i.; a (1, 2, 3) g.i. a *normalized* g.i. or *weak* g.i. Cline [9] calls a (1, 2, 4) g.i. a *left weak* g.i. and a (1, 2, 3) g.i. a *right weak* g.i. Notation is also variable but we shall reserve A^+ for the Moore–Penrose g.i. of the previous section. References to some applications of these g.i.'s can be found in [41].

The first point that we can make is that, as we remarked earlier, the only property used in the proof of Theorem 6.4 is (6.2.1), so that the result, and its corollary (i), page 134, still hold if A^+ is replaced by $A^{(1)}$, any member of $A(1)$, and C^+ by $C^{(1)}$. We can characterize $A(1)$ as follows [42]:

THEOREM 6.8 If $A^{(1)}$ is a (1) g.i. of A then all other members of $A(1)$ are given by

$$A^{(1)} + Y - A^{(1)}AYAA^{(1)} \tag{6.2.5}$$

as Y varies over all possible $n \times m$ matrices.

Proof. First, substituting (6.2.5) into (6.2.1) gives

$$AA^{(1)}A + AYA - AA^{(1)}AYAA^{(1)}A = A + AYA - AYA = A,$$

so that the matrix in (6.2.5) is certainly a (1) g.i. of A. Furthermore, if B is any other member of $A(1)$, then taking $Y = B - A^{(1)}$ in (6.2.5) produces

$$\begin{aligned} A^{(1)} + B - A^{(1)} - A^{(1)}A(B - A^{(1)})AA^{(1)} &\\ = B - A^{(1)}ABAA^{(1)} + A^{(1)}AA^{(1)}AA^{(1)} &\\ = B, & \end{aligned}$$

as required.

Using (6.2.5) it easily follows that if $b \neq 0$ then $x = A^{(1)}b$ ranges over all vectors satisfying (6.1.1) as $A^{(1)}$ ranges over all members of $A(1)$, provided the equations are consistent (i.e. $AA^{(1)}b = b$).

Another characterization [42] of $A(1)$ begins by rearranging rows and columns of A (if necessary) so that A may be assumed in the form (6.1.21). Then

THEOREM 6.9

$$A^{(1)} = \begin{bmatrix} G_1 & G_2 \\ G_3 & G_4 \end{bmatrix} \tag{6.2.6}$$

where $G_1 = A_1^{-1} - A_1^{-1}A_2G_3 - G_2A_3A_1^{-1} - A_1^{-1}A_2G_4A_3A_1^{-1}$ and G_2, G_3, G_4 are arbitrary matrices having appropriate dimensions. Also, rank $A \leqslant$ rank $A^{(1)} \leqslant \min(m, n)$.

We can deduce two interesting points from this theorem. First, if A is Hermitian and singular it does not necessarily follow that $A^{(1)}$ will be Hermitian, since for this to be the case we must have $G_2 = G_3^*$ and $G_4 = G_4^*$. Thus some but not all (1) g.i.'s of a Hermitian matrix are Hermitian, and conversely a square non-Hermitian matrix may have Hermitian (1) g.i.'s. Secondly, the condition on rank $A^{(1)}$ implies that a square singular matrix has nonsingular (1) g.i.'s.

At this stage it is interesting to examine commutativity of a square matrix and a g.i. Of course a nonsingular matrix always commutes with its inverse, but this is not necessarily true for singular matrices.

THEOREM 6.10 [31] A commutes with A^+ if and only if A^+ can be expressed as a polynomial in A (with scalar coefficients).

Proof. If $A^+ = f(A)$, clearly $AA^+ = A^+A$.

Conversely, if $AA^+ = A^+A$ we show that A^+ commutes with every matrix B which commutes with A, which establishes that A^+ is a polynomial in A [10, p. 136]. Using (6.2.1), (6.2.2) and $AB = BA$ we have

$$\begin{aligned} A^+B &= A^+AA^+B = (A^+)^2AB = (A^+)^2BA \\ &= (A^+)^2BA^3(A^+)^2 = (A^+)^2A^3B(A^+)^2 = AB(A^+)^2 \\ &= BA(A^+)^2 = BA^+. \end{aligned}$$

The only properties used in the preceding proof are (6.2.1) and (6.2.2), and in fact Hearon [20] has shown by a quite different method that if $X \in A(1, 2)$ commutes with A then X is a unique polynomial in A, and that a necessary and sufficient condition for the existence of such a matrix X is rank A = rank A^2 (see also [14] and Exercise 6.2.11).

An alternative necessary and sufficient condition [31] for A and A^+ to commute is that A be an *EP matrix*, defined by the condition $AZ = 0 \Leftrightarrow A^*Z = 0$, where Z is square. In particular if A is *EP* and has rank r it is called an *EPr* matrix. For other results involving *EPr* matrices and g.i.'s see [19], [21], [25] and [28].

We have remarked (Exercise 6.1.9) that the Moore–Penrose g.i. of an arbitrary matrix can be expressed in terms of the g.i. of a positive semidefinite Hermitian matrix. This also applies for other g.i.'s (see Exercise 6.2.6), and the following result is then useful.

THEOREM 6.11 [37] If the Hermitian matrix

$$A = \begin{bmatrix} A_1 & A_2 \\ A_2^* & A_3 \end{bmatrix}$$

is positive semidefinite, then

$$B = \begin{bmatrix} A_1^{(g)} + A_1^{(g)}A_2A_4^{(g)}A_2^*A_1^{(g)} & -A_1^{(g)}A_2A_4^{(g)} \\ -A_4^{(g)}A_2^*A_1^{(g)} & A_4^{(g)} \end{bmatrix},$$

where $A_4 = A_3 - A_2^*A_1^{(g)}A_2$,

(i) is a (1) g.i. of A if $(g) = (1)$,
(ii) is a (1, 2) g.i. of A if $(g) = (1, 2)$.

Further, if A_3 is nonsingular and rank A = rank A_1 + rank A_3, then

(iii) B is a (1, 2, 3) g.i. of A if $(g) = (1, 2, 3)$,
(iv) $B = A^+$ if $A_1^{(g)} = A_1^+$ and $A_4^{(g)} = A_4^+$.

For similar characterizations of g.i.'s of a general matrix see [30] and [40].

We can now give some analogues of Theorem 6.5. The proofs can be found in [23].

THEOREM 6.12 If (6.1.10) is consistent, then the solution having minimum norm is $A^{(1,4)}BC^{(1,3)}$.

Two corollaries of this result are of special interest.

THEOREM 6.13 For a given matrix A,

(i) the solution of $A^*AX = A^*$ having minimum norm is A^+,
(ii) the solution of $AXA = A$ having minimum norm is A^+.

For the system of linear equations (6.1.1) we can go further:

THEOREM 6.14

(i) If (6.1.1) is consistent, then $x = A^{(1,4)}b$ is the solution having minimum norm. Conversely, if Bb is the solution having minimum norm for every b such that (6.1.1) is consistent, then $B \in A(1, 4)$.
(ii) $x = A^{(1,3)}b$ is a least squares solution of (6.1.1). Conversely, if $x = Bb$ is a least squares solution for every b, then $B \in A(1, 3)$.

(iii) $x = A^+b$ is the least squares solution of (6.1.1) having minimum norm. Conversely, if $x = Bb$ is the least squares solution having minimum norm for every b, then $B = A^+$.

We next turn to the question of characteristic roots and vectors of a g.i. Following [38], we shall say that a g.i. B of a square matrix A has *property V* if: λ ($\neq 0$) is a characteristic root of A with corresponding characteristic vector x implies that $1/\lambda$ is a root of B with vector x, and conversely. Of course when A is nonsingular A^{-1} possesses property V, but it is helpful to define a weaker property: B has *property R* if reciprocals of nonzero characteristic roots of A are roots of B, and conversely. In general g.i.'s of various types need not possess either property, but Rohde [38] has proved the following two results.

THEOREM 6.15 If A is Hermitian then $A^{(1,3)}$ possesses property R.

Proof. For convenience write B to stand for $A^{(1,3)}$ and let λ be any nonzero characteristic root of A with $Ax = \lambda x$. Then

$$ABAx = \lambda ABx \Rightarrow Ax = \lambda ABx \Rightarrow x = ABx.$$

Since $(AB)^* = AB$, this gives

$$B^*Ax = x \Rightarrow \lambda B^*x = x \Rightarrow B^*x = (1/\lambda)x,$$

showing that B^* has a characteristic root $1/\lambda$, and therefore that B has a root $1/\lambda$, since λ is real.

Conversely, if μ is a nonzero characteristic root of B with characteristic vector y, then

$$\begin{aligned} By = \mu y &\Rightarrow ABy = \mu Ay \\ &\Rightarrow B^*Ay = \mu Ay \\ &\Rightarrow AB^*Ay = \mu A(Ay) \\ &\Rightarrow (1/\mu)Ay = A(Ay), \end{aligned}$$

showing that $1/\mu$ is a characteristic root of A.

THEOREM 6.16 If A is normal, then A^+ has property V. Conversely, if A and a (1) g.i. B of A are normal and B possesses property V, then $B = A^+$.

The proof follows using Exercise 6.2.12 and $A^+A = AA^+$ ((viii), Theorem 6.2).

Scroggs and Odell [39] have given a different definition of a unique g.i. A^S of a square matrix which possesses property V. They define $A^S = P^{-1}J^+P$, where J is the Jordan form of A, and P in the transformation $A = P^{-1}JP$ is chosen in a particular way so as to make A^S unique. It is easy to show that $A^S \in A(1, 2)$ and that if A is unitarily similar to its Jordan form (for example if A is normal) then A^S coincides with A^+. Drazin [13] has defined yet another g.i. A^D as the unique solution of the equations $AX = XA$, $X = X^2A$, $A^k = A^{k+1}X$, where k is the smallest positive integer such that rank A^k = rank A^{k+1}. A^D possesses property V but does not in general satisfy (6.2.1). Cline [9] gives a number of results involving the Drazin inverse. When A is an EP matrix both A^S and A^D coincide with A^+ [25].

We close this section by considering expressions for the derivative of a matrix which is a differentiable function of t, corresponding to the nonsingular case when

$$d(A^{-1})/dt = -A^{-1}\dot{A}A^{-1}. \tag{6.2.7}$$

THEOREM 6.17 [22] If $A(t)$ is differentiable and $B_1(t)$ is any differentiable (1) g.i. of A then $E_1\dot{A}E_2 = -A\dot{B}_1A$, where $E_1 = AB_1$ and $E_2 = B_1A$.

If $B_2(t)$ is any differentiable (1, 2) g.i. of A then $D_2\dot{B}_2D_1 = -B_2\dot{A}B_2$ where $D_1 = AB_2$, $D_2 = B_2A$.

Proof. Since $A = E_1A$ we have (see Exercise 6.2.13)

$$E_1\dot{E}_1A = E_1\dot{E}_1E_1A = 0.$$

Also,

$$\begin{aligned} E_1\dot{E}_1A &= E_1(\dot{A}B_1 + A\dot{B}_1)A \\ &= E_1\dot{A}E_2 + A\dot{B}_1A, \end{aligned}$$

giving

$$E_1\dot{A}E_2 = -A\dot{B}_1A. \tag{6.2.8}$$

If $B_2 \in A(1, 2)$ then (6.2.8) can be written

$$D_1\dot{A}D_2 = -A\dot{B}_2A. \tag{6.2.9}$$

Since $D_2B_2 = B_2 = B_2D_1$, multiplying (6.2.9) on the left and right by B_2 gives the desired result.

Hearon and Evans [22] give necessary and sufficient conditions for the existence of a differentiable g.i. (which we have implicitly assumed in the preceding theorem).

We have given only a few results involving different g.i.'s, corresponding to the most important properties of the ordinary inverse; many others can be found in the papers we have referred to.

EXERCISES

6.2.1 [19] If B_1 and B_2 are any two (1) g.i.'s of A, show that $B_1AB_2 \in A(1, 2)$.

6.2.2 If $B = A^{(1)}A$, show (i) B is idempotent, (ii) trace B = rank B = rank A, (iii) a general solution of the consistent equations $AX = Y$ is $A^{(1)}Y + (B - I_n)Z$, where Z is an arbitrary matrix having the same dimensions as X.

6.2.3 Show that rank $A^{(1)} \geqslant$ rank A and hence deduce that rank $A =$ rank $A^{(1)}$ if $A^{(1)} \in A(1, 2)$.

The converse also holds [38].

6.2.4 If A is factorized as in equation (6.1.4), show that B is a (1) g.i. of A if and only if $DBC = I_r$.

6.2.5 [31] Let $A^p + m_1A^{p+1} + \cdots + m_kA^{p+k}$ be the minimum polynomial of a square matrix A. If X is a (1, 2) g.i. of A and $AX = XA$, show by multiplying the minimum polynomial by X that $p = 0$ or 1. Hence derive an alternative proof of the necessity of the condition in Theorem 6.10.

6.2.6 Prove that $BAA^* = CAA^*$ implies $BA = CA$ by showing that $\|BA - CA\|^2 = 0$. Hence show that $A = AA^*(AA^*)^{(1)}A$.

Using this result show that if $A^{(1)}$ is any (1) g.i. of A, then (i) $A^{(1)}AA^{(1)} \in A(1, 2)$, and (ii) $A^*(AA^*)^{(1)} \in A(1, 2, 4)$ [41].

It can be shown in a similar fashion that [16]

$$(A^*A)^{(1)}A^* \in A(1, 2, 3)$$

and that [12]

$$A^*(AA^*)^{(1)}A(A^*A)^{(1)}A^* = A^+.$$

See also [24] and [21] for related results.

6.2.7 Show that $A^*(A^*AA^*)^{(1)}A^* = A^+$ by using the first part of the preceding exercise and Exercise 6.1.6.

See [43] for a different proof.

6.2.8 Using elementary transformations any $m \times n$ matrix A of rank r can be written as

$$PAQ = \begin{bmatrix} B & 0 \\ 0 & 0 \end{bmatrix}$$

where B is $r \times r$ and nonsingular. Let X be an $n \times m$ matrix such that

$$X = Q\begin{bmatrix} Z & U \\ V & W \end{bmatrix}P.$$

Show that X is a (1) g.i. of A if and only if $Z = B^{-1}$ and is a (1, 2) g.i. if and only if $Z = B^{-1}$ and $W = VBU$.

Conditions for X to be a (1, 2, 3) g.i. or to be the Moore–Penrose g.i. can be found in [30].

6.2.9 [40] If in the previous exercise P and Q are taken to be unitary, show that $X \in A(1, 2, 3)$ if and only if $Z = B^{-1}$, $W = 0$ and $U = 0$ and that $X = A^+$ if and only if $Z = B^{-1}$, $W = U = V = 0$.

6.2.10 If $B \in A(1)$ and B and A^r commute for some positive integer r, show that B possesses property R.

This result is based on one given in [34].

6.2.11 Englefield [14] calls a (1) g.i. of a square matrix A which commutes with A a *commuting inverse*. Show that if A_1 and A_2 are any two commuting inverses of A then $AA_1 = AA_2$. Hence show that if there is a commuting inverse which belongs to $A(1, 2)$ then it is unique (and is called the *commuting reciprocal* inverse [14] or *group* inverse [15].)

6.2.12 [14, 38] Show that the commuting reciprocal inverse of A (if it exists) has property V.

Englefield gives a number of other interesting results on the commuting and reciprocal inverses. Notice that when the group inverse exists (i.e. rank A = rank A^2) then it coincides with the Drazin inverse.

6.2.13 [22]. If $A(t)$ is differentiable, $B(t)$ is a differentiable (1) g.i. of A and $E = AB$, show that $E\dot{E}E = 0$.

6.2.14 [22]. In Theorem 6.17, show that $\dot{B}_2 = -B_2\dot{A}B_2$ if and only if $\dot{D}_2B_2 = B_2\dot{D}_1 = 0$ (compare with equation (6.2.7)).

6.3 Application to linear programming

Following [3], consider the linear programming problem: choose the column n-vector x so as to maximize

$$z = cx \tag{6.3.1}$$

subject to

$$a \leqslant Ax \leqslant b, \tag{6.3.2}$$

where A is an $m \times n$ matrix, a and b are column m-vectors and c a row n-vector (all taken to be real). We assume that the problem is feasible (i.e. that there is at least one vector satisfying (6.3.2)) and that the optimal value of z is finite. We also make the assumption that

$$\text{rank } A = m. \tag{6.3.3}$$

Let $A^{(1)}$, a member of $A(1)$, have columns $g_1, g_2, \ldots, g_m$. We then have an explicit solution of this class of linear programming problems:

THEOREM 6.18 [3] Optimal solutions of the problem defined by (6.3.1) and (6.3.2) are given by

$$x = \sum_{i\in J} g_i a_i + \sum_{i\in K} g_i b_i + \sum_{i\in L} g_i[\theta_i b_i + (1-\theta_i)a_i] + (I_n - A^{(1)}A)y, \tag{6.3.4}$$

where $J = \{i \mid cg_i < 0\}$, $K = \{i \mid cg_i > 0\}$, $L = \{i \mid cg_i = 0\}$, $0 \leqslant \theta_i \leqslant 1$ and y is an arbitrary column n-vector.

Proof. If for any column m-vector u we set

$$u = Ax \tag{6.3.5}$$

then (as mentioned on page 137) by virtue of (6.3.3) and (6.1.20) we can write (6.3.5) in the form

$$x = A^{(1)}u + (I_n - A^{(1)}A)y. \tag{6.3.6}$$

Substituting (6.3.5) and (6.3.6) into (6.3.1) and (6.3.2) produces the equivalent linear programming problem:

$$\begin{aligned} &\text{maximize} \quad cA^{(1)}u \\ &\text{subject to} \quad a \leqslant u \leqslant b. \end{aligned} \tag{6.3.7}$$

(The assumption that the original problem has a bounded solution implies that $c(I_n - A^{(1)}A)y = 0$ [3].) Since (6.3.7) gives $\sum_i (cg_i)u_i$ the optimal solution to this new problem is clearly

$$u_i = \begin{cases} a_i, & i \in J \\ b_i, & i \in K \\ \theta_i b_i + (1 - \theta_i)\, a_i, & i \in L, \end{cases}$$

and the desired expression is then obtained from (6.3.6).

It can be shown that the set (6.3.4) is independent of the particular $A^{(1)}$ involved.

Most linear programming problems arising in practice are feasible and have bounded optimal solutions, but the third assumption (6.3.3) is rather restrictive. In particular it is violated by problems with constraints $Ax \leqslant b$, $x \geqslant 0$, but Ben-Israel and Charnes [3] discuss ways of avoiding this difficulty by partitioning.

Theorem 6.18 has been used to obtain an iterative method for solving the problem of (6.3.1) and (6.3.2), this method also being applicable to general linear programming problems [36].

EXERCISE

6.3.1 If an $m \times n$ matrix A has left inverse B, and C is a matrix such that $(BC)^2 = \alpha BC$ for some scalar α, show that a left inverse of $A + \beta C$ is $B(I_m + \gamma CB)$ where $\gamma = -\beta/(1 + \alpha\beta)$ provided $\beta \neq -1/\alpha$.

This result has found application in linear programming theory [7].

REFERENCES

1. BEN-ISRAEL, A., 'A note on an iterative method for generalized inversion of matrices', *Math. of Comp.* **20,** 439–440 (1966).
2. BEN-ISRAEL, A. and CHARNES, A., 'Contributions to the theory of generalized inverses', *SIAM J. Appl. Math.* **11,** 667–699 (1963).
3. BEN-ISRAEL, A. and CHARNES, A., 'An explicit solution of a special class of linear programming problems', *Opns Res.* **16,** 1166–1175 (1968).

4. Ben-Israel, A. and Cohen, D., 'On iterative computation of generalized inverses and associated projections', *SIAM J. Numer. Anal.* **3**, 410–419 (1966).
5. Bjerhammar, A., 'Application of calculus of matrices to method of least squares, with special reference to geodetic calculations', *Trans. Roy. Inst. Tech. Stockholm*, **49**, 1–86 (1951).
6. Bjerhammar, A., 'Rectangular reciprocal matrices with special reference to geodetic calculations', *Bull. Géodésique*, **19**, 188–220 (1951).
7. Charnes, A. and Cooper, W. W., 'Structural sensitivity analysis in linear programming and an exact product form left inverse', *Naval Res. Logist. Quart.* **15**, 517–522 (1968).
8. Cline, R. E., 'Representations for the generalized inverse of a partitioned matrix', *SIAM J. Appl. Math.* **12**, 588–600 (1964).
9. Cline, R. E., 'Inverses of rank invariant powers of a matrix', *SIAM J. Numer. Anal.* **5**, 182–197 (1968).
10. Cullen, C. G., *Matrices and linear transformations*, Addison-Wesley, Reading, Massachusetts (1966).
11. Decell, H. P., Jr., 'An application of the Cayley–Hamilton Theorem to generalized matrix inversion', *SIAM Review*, **7**, 526–528 (1965).
12. Decell, H. P., Jr., 'An alternate form of the generalized inverse of an arbitrary complex matrix', *SIAM Review*, **7**, 356–358 (1965).
13. Drazin, M. P., 'Pseudo-inverses in associative rings and semi-groups', *Amer. Math. Monthly*, **65**, 506–514 (1958).
14. Englefield, M. J., 'The commuting inverses of a square matrix', *Proc. Cambridge Philos. Soc.* **62**, 667–671 (1966).
15. Erdelyi, I., 'On the matrix equation $Ax = \lambda Bx$', *J. Math. Anal. Appl.* **17**, 119–132 (1967).
16. Fisher, A. G., 'On construction and properties of the generalized inverse', *SIAM J. Appl. Math.* **15**, 269–272 (1967).
17. Greville, T. N. E., 'Some applications of the pseudoinverse of a matrix', *SIAM Review*, **2**, 15–22 (1960).
18. Greville, T. N. E., 'Note on the generalized inverse of a matrix product', *SIAM Review*, **8**, 518–521 (1966); **9**, 249 (1967).
19. Hearon, J. Z., 'Construction of *EPr* generalized inverses by inversion of nonsingular matrices', *J. Res. NBS*, **71B**, 57–60 (1967).
20. Hearon, J. Z., 'A generalized matrix version of Rennie's inequality', *J. Res. NBS*, **71B**, 61–64 (1967).
21. Hearon, J. Z. and Evans, J. W., 'On spaces and maps of generalized inverses', *J. Res. NBS*, **72B**, 103–107 (1968).

22. HEARON, J. Z. and EVANS, J. W., 'Differentiable generalized inverses', *J. Res. NBS*, **72B**, 109–113 (1968).
23. HEARON, J. Z., 'Generalized inverses and solutions of linear equations', *J. Res. NBS*, **72B**, 303–308 (1968).
24. KALMAN, R. E., 'New methods in Wiener filtering theory', *Proceedings of the first symposium on engineering applications of random function theory and probability*, 270–388, Wiley, New York (1963).
25. LEWIS, T. O. and NEWMAN, T. G., 'Pseudoinverses of positive semidefinite matrices', *SIAM J. Appl. Math.* **16**, 701–703 (1968).
26. MALIK, H. J., 'A note on generalized inverses', *Naval Res. Logist. Quart.* **15**, 605–612 (1968).
27. MAYNE, D. Q., 'An algorithm for the calculation of the pseudoinverse of a singular matrix', *Computer J.* **9**, 312–317 (1966).
28. MEYER, C. D., 'Some remarks on *EPr* matrices and generalized inverses', *Linear Algebra and its Applications*, **3**, 275–278 (1970).
29. MOORE, E. H., 'General Analysis, Part I', *Mem. Amer. Philos. Soc.* **1**, 197–209 (1935).
30. MORRISS, G. L. and ODELL, P. L., 'A characterization for generalized inverses of matrices', *SIAM Review*, **10**, 208–211 (1968).
31. PEARL, M. H., 'On generalized inverses of matrices', *Proc. Cambridge Philos. Soc.* **62**, 673–677 (1966).
32. PENROSE, R., 'A generalized inverse for matrices', *Proc. Cambridge Philos. Soc.* **51**, 406–413 (1955).
33. PENROSE, R., 'On best approximate solutions of linear matrix equations', *Proc. Cambridge Philos. Soc.* **52**, 17–19 (1956).
34. PRICE, C. M., 'The matrix pseudoinverse and minimal variance estimates', *SIAM Review*, **6**, 115–120 (1964).
35. RADO, R., 'Note on generalized inverses of matrices', *Proc. Cambridge Philos. Soc.* **52**, 600–601 (1956).
36. ROBERS, P. D. and BEN-ISRAEL, A., 'A suboptimization method for interval linear programming: a new method for linear programming', *Linear Algebra and its Applications*, **3**, 383–405 (1970).
37. ROHDE, C. A., 'Generalized inverses of partitioned matrices', *SIAM J. Appl. Math.* **13**, 1033–1035 (1965).
38. ROHDE, C. A., 'Some results on generalized inverses', *SIAM Review*, **8**, 201–205 (1966).
39. SCROGGS, J. E. and ODELL, P. L., 'An alternative definition of a pseudoinverse of a matrix', *SIAM J. Appl. Math.* **14**, 796–810 (1966).
40. TEWARSON, R. P., 'On some representations of generalized inverses', *SIAM Review*, **11**, 272–276 (1969).

41. Urquhart, N. S., 'Computation of generalized inverse matrices which satisfy specified conditions', *SIAM Review*, **10,** 216–218 (1968).
42. Urquhart, N. S., 'The nature of the lack of uniqueness of generalized inverse matrices', *SIAM Review*, **11,** 268–271 (1969).
43. Zlobec, S., 'An explicit form of the Moore–Penrose inverse of an arbitrary complex matrix', *SIAM Review*, **12,** 132–134 (1970).

7

Unimodular Matrices

7.1 Solution of linear equations in integers

The practical relevance of the material in this chapter is mainly concerned with linear programming problems of the type set out at the beginning of Section 1.5, with the additional requirement that a specified number of the components of x are allowed to take only integer values. Although the material we shall be considering here is not directly related to the control field, nevertheless we shall see that there are interesting correspondences in the underlying matrix theory. The particular way in which the integer programming problem (as it is usually known) is posed can vary somewhat—for example, the components of x may be restricted to be 0 or 1—but the difficulties involved in solving such problems are similar, and applications are both numerous and widespread [1]. Generally, excluding any fortunate coincidence, standard methods for solving linear programming problems will not produce a solution which also satisfies any integer requirements, and much effort has been devoted in recent years to devising special algorithms for solving integer programming problems. We do not intend to say any more about these as details are readily available in textbooks; see also [3] for an excellent and comprehensive survey with an extensive bibliography, and [11, 12] for important work on the structure of integer programming problems.

Because of the complexity of these special techniques, a pertinent question to ask is whether there are any classes of linear programming problems whose solutions will of necessity satisfy the integral conditions. It should be clear that to make this question meaningful we shall have to restrict attention to situations where *all* the variables must be integral, and that this condition will be independent of the values of the coefficients in the linear objective function (1.5.1).

Consider first the system of m linear equations

$$Ax = b \tag{7.1.1}$$

in n unknowns $x_1, x_2, \ldots, x_n$. If all the elements of A and b are integers we seek conditions for the solution of (7.1.1) also to be in integers for all possible m-vectors b. This problem was in fact solved in a classic paper by H. J. S. Smith in 1861 [23], but the methods he used are equally applicable to polynomial matrices (as we remarked on page 4—see also Appendix 1) and it is in the latter context that his work is usually referred to. We shall call a matrix having all its elements integral an *integer matrix*, and the necessary and sufficient condition for a square integer matrix to be invertible (i.e. have an inverse which is also an integer matrix) is that its determinant be ± 1. For completeness we give the result corresponding to Theorems 1.3 and 1.5:

THEOREM 7.1 Any $m \times n$ integer matrix A having rank r can be expressed as

$$A = PSQ \tag{7.1.2}$$

where P and Q are invertible integer matrices and

$$S = \text{diag}\,[i_1, i_2, \ldots, i_r, 0, \ldots, 0]$$

is the *Smith form* of A. Each integer i_j divides i_{j+1}, and the specific form of S depends on the relative magnitudes of r, m and n.

Using the preceding theorem we can now demonstrate Smith's result for (7.1.1), in which we assume $m \leqslant n$.

THEOREM 7.2 The equations (7.1.1) have an integral solution for all possible integral b if and only if the Smith form of A is $[I_m\ 0]$.

Proof. From (7.1.1) and (7.1.2) we have

$$Sy = c \tag{7.1.3}$$

where $y = Qx$ and $c = P^{-1}b$. Since P and Q are invertible x will be integral if and only if y is integral, and similarly for b and c. The equations (7.1.3) have an integral solution for all possible c_i if and only if s_i divides c_i $(i = 1, 2, \ldots, m)$ and this can only be true if all the s_i are unity.

If rank $A < m$ then the equations will not be consistent for all possible b.

Smith called an integer matrix satisfying the condition of Theorem 7.2 a *prime* matrix, and the similarity with the definition of relatively prime polynomial matrices on page 16 is apparent. It is also interesting that when $m = 1$ and $n = 2$, Theorem 7.2 simply states that a necessary and sufficient condition for the single equation $a_{11}x_1 + a_{12}x_2 = b_1$ to have a solution in integers is that a_{11} and a_{12} be relatively prime, and this is the analogue of the result which we stated for polynomials at the beginning of Section 1.2.

It is important to realize that Theorem 7.2 only gives a condition for the existence of integral solutions of (7.1.1), *not* for all solutions to be integral. This is best illustrated by a simple example:

$$\begin{bmatrix} 1 & 0 & 4 \\ 0 & 1 & 2 \end{bmatrix} \begin{bmatrix} x_1 \\ x_2 \\ x_3 \end{bmatrix} = \begin{bmatrix} b_1 \\ b_2 \end{bmatrix}.$$

Clearly $S = [I_2 \; 0]$ and the general solution can be written as $x_1 = b_1 - 4t$, $x_2 = b_2 - 2t$, $x_3 = t$ (t an arbitrary integer). If x_1 or x_2 is zero this will not be integral for all possible b_1 and b_2.

If A has rank m then a set of m linearly dependent columns of A is called a *basis* and the corresponding x in (7.1.1), having m nonzero components, is called a *basic* solution. For application to integer programming Theorem 7.2 is not adequate, since we require all basic solutions to be integral so that the optimal solution will be integral, whatever the values of the coefficients in the objective function. A stronger condition on A is needed, namely that all nonzero $m \times m$ minors of A have value ± 1; in this case A is called *unimodular* and we have

THEOREM 7.3 All basic solutions of (7.1.1) are integral for all integral b if and only if A is unimodular.

Proof [25]. Let B be any nonsingular $m \times m$ submatrix of A. If A is unimodular then B is invertible, so the corresponding components of x will be integral.

Conversely, if all basic solutions are integral then in particular there exist vectors y and z each having m nonzero integer components satisfying $By = b$ and $Bz = b + e_i$ where e_i is the ith column of I_m. Hence $B(z - y) = e_i$, or $z - y = B^{-1}e_i$, showing that the ith column of B^{-1} is integral. Since this holds for $i = 1, 2, \ldots, m$ it follows that B^{-1} is an integer matrix which implies that $\det B = \pm 1$.

Some relaxation of the condition in Theorem 7.3 can be achieved by proceeding as follows. Consider a set S of m-vectors having integer components, and let a basis in S be $s_1, s_2, \ldots, s_m$. If d is any vector in S we say that S is a *unimodular set* if, in

$$d = \sum_{i=1}^{m} \lambda_i s_i,$$

the coefficients λ_i are all 0, $+1$ or -1, for every basis and every d in S (S is also sometimes said to have the *Dantzig* property [15]). In particular we are interested in the case where S is the set of columns of the matrix A, but it is important to realize that the definitions of a unimodular matrix and of its columns forming a unimodular set are *not* equivalent.

This is apparent from a simple example:

$$A = \begin{bmatrix} 1 & 4 & 5 \\ 2 & 5 & 7 \end{bmatrix} = [a_1, a_2, a_3].$$

Obviously $a_3 = a_1 + a_2$, so the set of columns of A is unimodular, but the matrix A is not unimodular, having 2×2 minors equal to ± 3. If we look at the corresponding linear equations

$$\begin{aligned} x_1 + 4x_2 + 5x_3 &= b_1 \\ 2x_1 + 5x_2 + 7x_3 &= b_2 \end{aligned} \tag{7.1.4}$$

it is easy to verify that all basic solutions are integral if and only if $2b_1 - b_2$ is divisible by 3. In general we have the result:

THEOREM 7.4 [18] All basic solutions of (7.1.1) are integral if and only if the set of columns of A is unimodular, provided b is such that a basic integral solution exists.

The vital difference between Theorems 7.3 and 7.4 is that if the columns of A form a unimodular set but A is not a unimodular matrix, then only for certain vectors b will all basic solutions of (7.1.1) be integral; otherwise no basic solutions are integral. We stress this point since it is not always made clear in the existing literature. The set of suitable vectors b is simply $\{\sum_{i=1}^{m} t_i a_i\}$, where $a_1, a_2, \ldots, a_m$ is any basis in A and $t_1, t_2, \ldots, t_m$ are arbitrary integers, this set being called the *integral span* of the columns of A. Thus if, for instance, in the preceding example (7.1.4) we take $b_1 = t_1 + 4t_2$, $b_2 = 2t_1 + 5t_2$, then the basic solutions are $x_1 = t_1$, $x_2 = t_2$, $x_3 = 0$; $x_1 = t_1 - t_2$, $x_2 = 0$, $x_3 = t_2$; and $x_1 = 0$, $x_2 = t_2 - t_1$, $x_3 = t_1$. Notice, however, that if A is unimodular then its columns do form a unimodular set. For if $a_1, \ldots, a_m$ is any basis in A and d is any other column of A we have

$$d = \sum_{i=1}^{m} x_i a_i \tag{7.1.5}$$

with all x_i integral by virtue of Theorem 7.3. If d and some vector a_j corresponding to any nonzero x_j are interchanged in (7.1.5) this gives a new basis in A in which

$$a_j = \left(d - \sum_{i \neq j} x_i a_i\right) \Big/ x_j. \tag{7.1.6}$$

Since A is unimodular the coefficients in (7.1.6) must be integers, so it follows that $x_j = \pm 1$.

Before moving to our final definition we can remark that in linear programming the solution of (7.1.1) is usually required to be non-negative, but this introduces no further restriction on A, since if all basic solutions are integral, nonnegative solutions (if they exist) will also be integral.

We end this section by considering the system of linear inequalities

$$Ax \leqslant b \tag{7.1.7}$$

which often occur in linear programming problems in place of (7.1.1). As before A is an $m \times n$ integer matrix and b an integer column m-vector, so we can write (7.1.7) in the equivalent form

$$y + Ax = b, \qquad y \geqslant 0 \tag{7.1.8a}$$

or

$$[I_m \quad A]\begin{bmatrix} y \\ x \end{bmatrix} = b, \qquad y \geqslant 0. \tag{7.1.8b}$$

We now require all basic solutions of (7.1.8) to be integral for all possible b. For example, if $x_1, y_2, y_3, \ldots, y_m$ are allowable as basic variables then clearly a_{11} must be ± 1 since $x_1 = b_1/a_{11}$. A similar argument shows that this must hold for all nonzero elements of A, and by choosing basic solutions containing $m - 2$, $m - 3, \ldots, 1$ components of y it can easily be seen that all minors of A of all orders must be 0, +1 or −1; in this case A is called *totally unimodular*. The condition we have just discussed was shown to be necessary *and* sufficient by Hoffman and Kruskal [21]:

THEOREM 7.5 All basic solutions of (7.1.7) (or equivalently (7.1.8)) are integral for all integral b if and only if A is totally unimodular.

Veinott and Dantzig [25] have pointed out that Theorem 7.5 follows easily from Theorem 7.3 by replacing A by $[I_m \; A]$. Notice also that $[I_m \; A]$ is prime (in Smith's sense) but although Theorem 7.2 then tells us that an integer solution exists this is in fact obvious (namely, $x = 0$, $y = b$). Again we remark that the nonnegative condition on y (and possibly also x) imposes no further restriction on A.

Other material on various unimodular properties can be found in [4], [16], [17] and [18] (see also Exercises 7.1.3–7.1.6), and some interesting relationships with combinatorial analysis are explored in [22], but it is to an important class of linear programming problems for which the constraint matrix is totally unimodular that we devote our attention in the next section.

EXERCISES

7.1.1 Find invertible matrices P and Q such that PAQ is in Smith form, where

$$A = \begin{bmatrix} 1 & 0 & -1 & 4 \\ 2 & -1 & 0 & 0 \\ 3 & -2 & 0 & -1 \end{bmatrix}.$$

Hence find the general solution in integers of the set of equations $Ax = b$, where b is an integer vector.

See [8] for a different approach.

7.1.2 [23] Show that a solution in integers of equation (7.1.1) exists if and only if the mth determinantal divisors of A and $[A, b]$ are equal.

7.1.3 [5] If A is totally unimodular and nonsingular show that A^{-1} is totally unimodular.

7.1.4 [6] Show that if two columns of a totally unimodular matrix are such that any zero elements appear in the same rows, then either the two columns are identical or one is the negative of the other.

Cederbaum [5, 6] gives a number of interesting results on totally unimodular matrices arising in the context of electrical network theory.

7.1.5 [7] By considering the equations $Ax = b$ where

$$b_i = \tfrac{1}{2} \sum_{j=1}^{n} a_{ij},\ i = 1, 2, \ldots, n,$$

show that if the row sums of an integer matrix A are even then $\det A$ is even.

7.1.6 [7] Using a method similar to that in the previous exercise, show that for any integer matrix A, $\det A$ is a multiple of the greatest common divisor of

$$\sum_{j\in T} \lambda_j a_{1j},\ \sum_{j\in T} \lambda_j a_{2j},\ \ldots,\ \sum_{j\in T} \lambda_j a_{nj}$$

where T is a subset of the columns of A and $\lambda_j = \pm 1$, all j.

This result can be used to give a necessary and sufficient condition for total unimodularity [7].

7.2 Transportation problems

The transportation problem in linear programming is very well known and fully described in the many textbooks on the subject (see, for example, [9]) so we confine ourselves here to a brief outline. The name arises from the situation of transporting goods from m (>1) points of origin to n (>1) destinations. If the amount available at origin i is a_i, the amount required at destination j is b_j and the unit cost of transportation from origin i to destination j is c_{ij}, the problem is to determine the amounts x_{ij} to be sent from origin i to destination j so as to satisfy the requirements at least possible cost. We can assume without loss of generality that the total amount available is equal to the total required, i.e.

$$\sum_{i=1}^{m} a_i = \sum_{j=1}^{n} b_j \tag{7.2.1}$$

and that a_i, b_j and c_{ij} are all integers. The mathematical formulation is:

$$\text{Minimize} \qquad \sum_{i=1}^{m} \sum_{j=1}^{n} c_{ij} x_{ij} \tag{7.2.2}$$

subject to

$$\sum_{j=1}^{n} x_{ij} = a_i, \qquad (i = 1, 2, \ldots, m) \tag{7.2.3}$$

$$\sum_{i=1}^{m} x_{ij} = b_j, \qquad (j = 1, 2, \ldots, n), \tag{7.2.4}$$

and of course for practical reasons all $x_{ij} \geqslant 0$. These equations can be represented diagrammatically as in Fig. 1, the row and column sums being a_i and b_j respectively.

Equations (7.2.3) and (7.2.4) can be combined together in the form

$$Ux = d \tag{7.2.5}$$

where $x = [x_{11}, x_{12}, \ldots, x_{1n}, x_{21}, x_{22}, \ldots, x_{2n}, \ldots, x_{mn}]^T$, $d = [a_1, a_2, \ldots, a_m, b_1, b_2, \ldots, b_n]^T$ and the $(m + n) \times (mn)$ constraint matrix is

$$U(m, n) = \overset{\leftarrow n \text{ cols} \rightarrow}{\begin{bmatrix} 1 & 1 & \cdot & \cdot & 1 & & & & & & & & & & \\ \cdot & \cdot & \cdot & \cdot & \cdot & 1 & 1 & \cdot & \cdot & 1 & & & & & \\ & & & & & & & & & \cdot & & & & & \\ \cdot & \cdot & & & \cdot & \cdot & \cdot & & \cdot & & \cdot & & & & \\ & & & & & & & & & & & \cdot & & & \\ \cdot & \cdot & & & \cdot & \cdot & \cdot & & \cdot & & & 1 & 1 & \cdot & \cdot & 1 \\ 1 & \cdot & & & \cdot & 1 & \cdot & & \cdot & & & 1 & & & & \cdot \\ & 1 & \cdot & & \cdot & & 1 & & \cdot & & & & 1 & & & \cdot \\ & & \cdot & & \cdot & & & \cdot & \cdot & & & & & & \cdot & \cdot \\ & & & & & & & \cdot & & & & & & & \cdot & \cdot \\ & & & \cdot & \cdot & & & & \cdot & & & & & & & \\ & & & & 1 & & & & 1 & & & & & & & 1 \end{bmatrix}} \begin{matrix} \uparrow \\ m \text{ rows} \\ \downarrow \\ \uparrow \\ n \text{ rows} \\ \downarrow \end{matrix} \;. \tag{7.2.6}$$

In terms of the columns e_i of I_{m+n}, a simple expression is

$$U(m, n) = [e_1 + e_{m+1}, e_1 + e_{m+2}, \ldots, e_1 + e_{m+n}, e_2 + e_{m+1}, \ldots, e_2 + e_{m+n}, \ldots, e_m + e_{m+1}, \ldots, e_m + e_{m+n}] \tag{7.2.7}$$

so the columns of $U(m, n)$ are

$$e_i + e_{m+j}, \quad (i = 1, 2, \ldots, m, \quad j = 1, 2, \ldots, n). \tag{7.2.8}$$

Notice that (7.2.1) implies that one of the equations in (7.2.5) is redundant, so $U(m, n)$ has rank $m + n - 1$. What is of particular interest to us here is that any set of $m + n - 1$ rows of $U(m, n)$ forms a unimodular matrix, so that all basic solutions of (7.2.5) are integral for all possible integral a_i and b_j. Algorithms for solving the transportation problem which take account of this special structure naturally prove more efficient than standard linear programming methods. In fact we shall see that $U(m, n)$ is itself totally unimodular, but we first show that

THEOREM 7.6 [9] Any nonsingular submatrix of $U(m, n)$ having order $m + n - 1$ can be expressed as a triangular matrix after suitable permutations of rows and columns. In consequence all basic solutions of (7.2.5) are integral.

Proof. Our presentation deals directly with the matrix $U(m, n)$ and so is slightly different from that of Dantzig [9, p. 303] who develops the argument in terms of the cells in Fig. 1.

	b_1	b_2	.	.	.	b_n
a_1	x_{11}	x_{12}	.	.	.	x_{1n}
a_2	x_{21}	x_{22}	.	.	.	x_{2n}
.	:	.	.	.	.	.
.	.	.	.	.	.	.
a_m	x_{m1}	x_{m2}	.	.	.	x_{mn}

Fig. 1

Any basis consists of $m + n - 1$ columns of $U(m, n)$ forming a matrix C, say, of order $(m + n) \times (m + n - 1)$. Then there must be some row of C containing a single unit element, all other entries being zero. For if not, then every row of C must contain at least two unit elements, which implies that in (7.2.8) each value of i and j occurs at least twice. Inspection of (7.2.7) shows that this requires that C has at least $2n$ or $2m$ columns whichever is the greater, which is impossible since $n + m - 1 < \max(2m, 2n)$.

Let C_1 be the matrix of order $(m + n - 1) \times (m + n - 2)$ obtained from C by deleting the row (i_1, say) and column (j_1, say) selected by the preceding process in which the single unit element lies. Then C_1 also has a row with just a single nonzero element, for the contrary assumption, that every row of C_1 contains at least two nonzero elements, again leads to a contradiction. To see this, we must consider two cases; first, if $i_1 \in \{1, 2, \ldots, m\}$ then the columns of C_1 are given by $f_i + f_{m-1+j}$, $i = 1, 2, \ldots, m - 1$, $j = 1, 2, \ldots, n$, where f_i is the ith column of I_{m+n-1}. The assumption on the rows of C_1 implies that C_1 has at least $2n$ or $2(m - 1)$ columns, whichever is the greater. Secondly, if $i_1 \in \{m + 1, \ldots, n\}$ then the columns of C_1 are given by $f_i + f_{m-j}$, $i = 1, \ldots, m$, $j = 1, \ldots, n - 1$ which implies that C_1 has at least $2m$ or $2(n - 1)$ columns, whichever is the greater. It is easy to verify that none of these situations is possible, so C_1 has some row i_2 which contains a single nonzero element in some column j_2. Deleting these gives a matrix C_2 and the argument then continues in the same way until all the columns of C have been chosen. Thus if the rows of C are selected in the sequence $i_1, i_2, \ldots$, and the columns according to $j_1, j_2, \ldots$, the resulting matrix is triangular.

This triangular matrix has determinant equal to 1 since all the elements on the principal diagonal are unity. Hence it follows that any nonsingular submatrix of U of order $m + n - 1$ has determinant equal to ± 1 and so by Theorem 7.3 all basic solutions of (7.2.5) are integral.

It is easy to show [9] that basic solutions exist which satisfy the requirement $x_{ij} \geqslant 0$, all i, j.

As stated earlier $U(m, n)$ is totally unimodular, but rather than show this directly we prove that a class of matrices, of which $U(m, n)$ is a member, is totally unimodular.

Let P be any matrix having the following properties:

($P1$) Every column of P contains at most two nonzero entries.
($P2$) Every element of P is either 0, $+1$ or -1.
($P3$) The rows of P can be separated into two disjoint sets R and S such that if two nonzero entries p_{ij}, p_{kj} in a column of P have the same sign then $i \in R$ and $k \in S$, but if p_{ij} and p_{kj} have opposite signs then either $i, k \in R$ or $i, k \in S$.

It is obvious from (7.2.7) that $U(m, n)$ is an example of the matrices just defined.

THEOREM 7.7 [15] Every matrix P having properties ($P1$), ($P2$) and ($P3$) is totally unimodular.

Proof. We follow Hoffman in an appendix to [15], and prove by induction that any $k \times k$ submatrix Q of P has determinant 0, $+1$ or -1. It is easy to verify that Q still satisfies the conditions ($P1$), ($P2$) and ($P3$), so in particular by ($P2$) the result holds for $k = 1$. Suppose that the assumption is true for $k - 1$. Then if every column of Q has two nonzero entries, it follows from ($P3$) that the rows q_i of Q are related by

$$\sum_{i \in R} q_i = \sum_{i \in S} q_i$$

which implies that $\det Q = 0$. If any column of Q has no nonzero elements then $\det Q = 0$. Finally, if some column of Q contains a single nonzero entry, expansion by this column gives $\det Q = \pm \det Q'$ where $\det Q'$ is the cofactor of the nonzero element and so is equal to 0, $+1$ or -1 by the induction hypothesis. Thus in all cases $\det Q$ is equal to 0, $+1$ or -1 and this establishes the theorem.

Gale has shown in the same appendix to [15] that if a matrix has at most two nonzero elements in each column then ($P2$) and ($P3$) are necessary conditions for it to be totally unimodular.

As we indicated previously, the structure of the transportation problem as revealed by the preceding theorems has resulted in the construction of simplified computational methods of solution. This has naturally led to a search for related problems which can be transformed into the transportation problem. This work is due to Heller; first [20] consider the problem of minimizing

$$ky \tag{7.2.9}$$

subject to

$$Zy = g, \qquad y \geqslant 0 \tag{7.2.10}$$

where Z has mn columns and rank $m + n - 1$. Comparing (7.2.9) and (7.2.10) with (7.2.2) and (7.2.5), we wish to find conditions for the existence of a nonsingular matrix T and a permutation matrix P such that $y = Px$, $TZP = U$, $Tg = d$ and $kP = c$ where $c = [c_{11}, c_{12}, \ldots, c_{1n}, \ldots, c_{mn}]$. Heller gives necessary and sufficient conditions in terms of the columns of Z for the existence of suitable matrices T and P, and refers to an algorithm (said to be simple) for determining when equivalence holds. In a second paper [19] Heller deals with a more general form of equivalence which does not involve the objective functions.

We now turn to a generalization of the transportation problem which has a number of interesting applications [13]. This is the so-called 'three-dimensional' or 'solid' transportation problem, the name arising since the situation can be thought of as a rectangular block in which the layers in any of the directions parallel to the faces form transportation problems of the previous ('two-dimensional') type (see Figs. 1 and 3). Specifically, for given integers $a_{ik}, b_{jk}, e_{ij}, c_{ijk}$ we wish to choose x_{ijk} so as to minimize

$$\sum_{i=1}^{m} \sum_{j=1}^{n} \sum_{k=1}^{p} c_{ijk} x_{ijk}$$

subject to

$$\sum_{j=1}^{n} x_{ijk} = a_{ik}, \qquad (i = 1, 2, \ldots, m;\ k = 1, 2, \ldots, p) \tag{7.2.11}$$

$$\sum_{i=1}^{m} x_{ijk} = b_{jk}, \qquad (j = 1, 2, \ldots, n;\ k = 1, 2, \ldots, p) \tag{7.2.12}$$

$$\sum_{k=1}^{p} x_{ijk} = e_{ij}, \qquad (i = 1, 2, \ldots, m;\ j = 1, 2, \ldots, n) \tag{7.2.13}$$

$(m, n, p > 1)$.

For example, when $m = 2$, $n = 3$ and $p = 3$ we can represent the equations in the pictorial form shown in Figs. 2 and 3. Each cell in Fig. 3 contains one of the x_{ijk}, which when summed in vertical columns

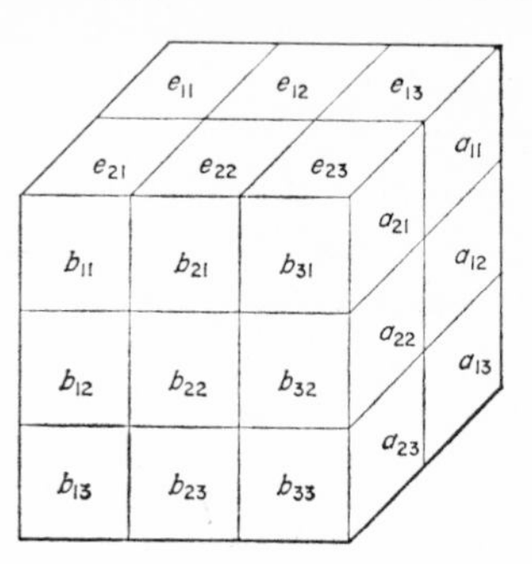

Fig. 2

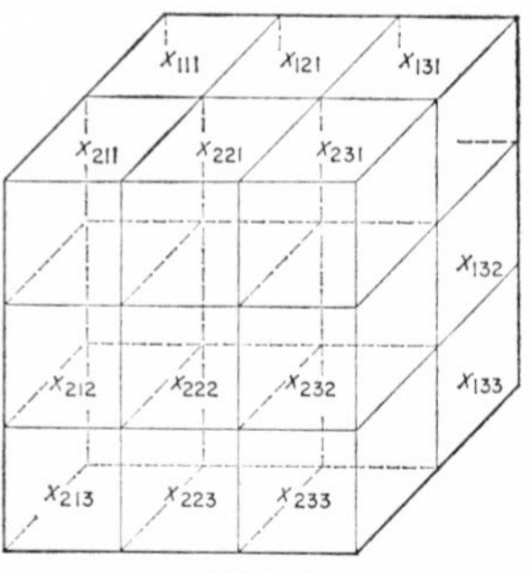

Fig. 3

total e_{ij}, and when summed in horizontal rows give a_{ik} and b_{jk}. The conditions corresponding to (7.2.1) for consistency of the equations are

$$\left.\begin{aligned} \sum_{i=1}^{m} a_{ik} = \sum_{j=1}^{n} b_{jk}, \quad \sum_{k=1}^{p} b_{jk} = \sum_{i=1}^{m} e_{ij}, \quad \sum_{j=1}^{n} e_{ij} = \sum_{k=1}^{p} a_{ik} \\ \sum_{i}\sum_{k} a_{ik} = \sum_{j}\sum_{k} b_{jk} = \sum_{i}\sum_{j} e_{ij}. \end{aligned}\right\} \tag{7.2.14}$$

Comparing (7.2.11) and (7.2.12) with the two-dimensional problem in (7.2.5) we see that the kth 'layer' of the solid problem can be written as

$$U(m, n)y_k^T = \begin{bmatrix} a_k^T \\ b_k^T \end{bmatrix}$$

where

$$y_k = [x_{11k}, x_{12k}, \ldots, x_{1nk}, x_{21k}, x_{22k}, \ldots, x_{2nk}, \ldots, x_{m1k}, \ldots, x_{mnk}],$$

$a_k = [a_{1k}, a_{2k}, \ldots, a_{mk}]$, $b_k = [b_{1k}, b_{2k}, \ldots, b_{nk}]$. Thus we can combine together equations (7.2.11), (7.2.12) and (7.2.13) in the form

$$Vz = f \tag{7.2.15}$$

where

$$z = [y_1, y_2, \ldots, y_p]^T, \quad f = [a_1, b_1, a_2, b_2, \ldots, a_p, b_p, e]^T,$$
$$e = [e_{11}, e_{12}, \ldots, e_{1n}, e_{21}, \ldots, e_{2n}, \ldots, e_{m1}, \ldots, e_{mn}]$$

and

$$V(m, n, p) = \begin{bmatrix} U(m,n) & & & \\ & U(m,n) & & \\ & & \ddots & \\ & & & U(m,n) \\ I_{mn} & I_{mn} & \cdots & I_{mn} \end{bmatrix}. \quad (7.2.16)$$

$\longleftarrow$ p blocks $\longrightarrow$

In (7.2.16), $V(m, n, p)$ has dimensions $(mn + np + pm) \times (mnp)$ and by (7.2.14) has rank $N(m, n, p) = mnp - (m - 1)(n - 1)(p - 1)$. A basic solution of (7.2.15) is one containing $N(m, n, p)$ non-zero variables, but unlike the standard transportation problem there need not necessarily exist basic solutions which also satisfy $x_{ijk} \geqslant 0$, all i, j, k. It is now natural to ask whether $V(m, n, p)$ has any of the properties of $U(m, n)$, and in particular whether it is totally unimodular. This last question can be immediately answered in the negative by constructing a simple counter-example (see Exercise 7.2.1). However, Haley [14] gives a method for constructing basic solutions in integers of (7.2.15) which suggests that all bases can be rearranged to give triangular matrices as in Theorem 7.6. This would imply that the columns of $V(m, n, p)$ form a unimodular set, but efforts by the author to establish this, including an attempt to extend the method of proof of Theorem 7.6, have so far not met with success except for the case $p = 2$. Nevertheless, it does seem promising to continue along these lines, so as to try to obtain classes of unimodular sets of vectors more general than that given by the columns of the matrix in Theorem 7.7. In particular this might give unimodular sets with more than two nonzero elements in some columns. Mathematical formulation of transportation problems having dimension greater than three is clearly straightforward and may also lead to yet further classes of unimodular sets.

EXERCISE

7.2.1. Construct the submatrix formed by rows 1, 4, 7, 8, 11, 15, 16 and columns 1, 3, 4, 11, 12, 13, 14 of $V(3, 3, 2)$ and verify that its determinant is equal to -2.

7.3 Combinatorial equivalence

If A is an $m \times n$ unimodular matrix (with $m \leqslant n$) then so is QA, where Q is any $m \times m$ invertible integer matrix—this follows directly from the fact that if B is any $m \times m$ submatrix of A then $\det QB = \pm \det B$. However, if P is an $n \times n$ invertible matrix clearly AP need not be unimodular, as the following simple example illustrates with A unimodular but AP having a second order minor equal to 2:

$$A = \begin{bmatrix} 1 & 0 & 1 \\ 0 & 1 & 1 \end{bmatrix}, \quad P = \begin{bmatrix} 1 & 3 & 0 \\ 0 & 1 & 0 \\ 0 & 0 & 1 \end{bmatrix}, \quad AP = \begin{bmatrix} 1 & 3 & 1 \\ 0 & 1 & 1 \end{bmatrix}.$$

Of course, if P is a permutation matrix then multiplication of A on the right by P simply rearranges the columns, so QAP *is* unimodular.

Our last remark leads us to a special type of equivalence relation for totally unimodular matrices, but before giving this we need the following definition. A *pivot operation* on any nonzero element a_{ij} of an $m \times n$ matrix A results in a matrix

$$A^{ij} = \frac{1}{a_{ij}} \begin{bmatrix} d_{11} & d_{12} & \cdots & -a_{1j} & \cdots & d_{1n} \\ d_{21} & d_{22} & \cdots & -a_{2j} & \cdots & d_{2n} \\ \cdot & \cdot & \cdots & \cdot & \cdots & \cdot \\ a_{i1} & a_{i2} & \cdots & 1 & \cdots & a_{in} \\ \cdot & \cdot & \cdots & \cdot & \cdots & \cdot \\ d_{m1} & d_{m2} & \cdots & -a_{mj} & \cdots & d_{mn} \end{bmatrix} \tag{7.3.1}$$

where

$$d_{rs} = \begin{vmatrix} a_{rs} & a_{rj} \\ a_{is} & a_{ij} \end{vmatrix}, \qquad r \neq i,\ s \neq j. \tag{7.3.2}$$

The significance of this operation is easily explained. For convenience write the equations (7.1.1) in the form

$$\left.\begin{aligned} v_1 + a_{11}x_1 + a_{12}x_2 + \cdots + a_{1n}x_n &= 0 \\ &\vdots \\ v_m + a_{m1}x_1 + a_{m2}x_2 + \cdots + a_{mn}x_n &= 0 \end{aligned}\right\} \tag{7.3.3}$$

where $v = -b$. If $a_{ij} \neq 0$ then the ith equation in (7.3.3) gives

$$x_j = -(v_i + a_{i1}x_1 + \cdots + a_{i,j-1}x_{j-1} + a_{i,j+1}x_{j+1} + \cdots + a_{in}x_n)/a_{ij} \tag{7.3.4}$$

and substitution of (7.3.4) into (7.3.3) produces

$$\left.\begin{aligned} v_1 + k_{11}x_1 + \cdots + k_{1j}v_i + \cdots + k_{1n}x_n &= 0 \\ &\vdots \\ x_j + k_{i1}x_1 + \cdots + k_{ij}v_i + \cdots + k_{in}x_n &= 0 \\ &\vdots \\ v_m + k_{m1}x_1 + \cdots + k_{mj}v_i + \cdots + k_{mn}x_n &= 0 \end{aligned}\right\} \tag{7.3.5}$$

where $k_{is} = a_{is}/a_{ij}$, $s \neq j$; $k_{rj} = -a_{rj}/a_{ij}$, $r \neq i$; $k_{ij} = 1/a_{ij}$; $k_{rs} = d_{rs}/a_{ij}$, $r \neq i$, $s \neq j$, d_{rs} being given by (7.3.2). In view of (7.3.1) we can write (7.3.3) and (7.3.5) respectively as

$$[I_m \quad A]\begin{bmatrix} v \\ x \end{bmatrix} = 0 \tag{7.3.6}$$

and

$$[I_m \quad A^{ij}]\begin{bmatrix} v' \\ x' \end{bmatrix} = 0 \tag{7.3.7}$$

where v' is obtained from v by replacing v_i by x_j and x' is obtained from x by replacing x_j by v_i. Thus a pivot operation interchanges the roles of x_j and v_i, and since it is clearly reversible the solutions of (7.3.6) and (7.3.7) must be identical. A basic solution of (7.3.3) in terms of m of the x_i can be obtained by a sequence of m pivot operations; in particular if $m = n$ a square nonsingular matrix A can be inverted in n pivot operations (see [10] for further details).

Any $m \times n$ matrix B is said to be *combinatorially equivalent* to A ($B :: A$) if and only if B can be obtained from A by a finite sequence of row permutations, column permutations and pivot operations on nonzero elements. An equivalent definition is that $B :: A$ if and only if there exist an $(m + n) \times (m + n)$ permutation matrix P and an $m \times m$ nonsingular matrix Q such that

$$Q[I_m \quad A]P = [I_m \quad B]. \tag{7.3.8}$$

There are at most $(m + n)!$ matrices which are combinatorially equivalent to A, and the maximum number is attained if and only if every set of m columns of $[I_m \; A]$ is linearly independent. The concept of combinatorial equivalence was first introduced by Tucker [24] who gave a number of theorems, and an account of some of these properties can also be found in the textbook [10], including a proof of the equivalence of the two definitions.

What is of interest to us here is the application of the preceding definitions to integer matrices. We see at once from (7.3.1) and (7.3.2) that if A is totally unimodular then A^{ij} will be an integer matrix, since all d_{rs} are 0, ± 1. It also follows in this case (see [10, p. 231]) that the matrix Q in (7.3.8) is an invertible integer matrix, and by comparing equations (7.1.8b) and (7.3.8) and using Theorem 7.5 we can deduce:

THEOREM 7.8 If A is totally unimodular so is any matrix which is combinatorially equivalent to A.

A more general form of pivoting can be defined by simultaneously interchanging more than one pair of variables in equations (7.3.3). Suppose that $x_1, x_2, \ldots, x_k$ and $v_1, v_2, \ldots, v_k$ are to be interchanged in equations (7.3.3), which we write here in the partitioned form

$$\begin{bmatrix} v^1 \\ v^2 \end{bmatrix} + \begin{bmatrix} A_{11} & A_{12} \\ A_{21} & A_{22} \end{bmatrix}\begin{bmatrix} x^1 \\ x^2 \end{bmatrix} = 0, \tag{7.3.9}$$

where $v^1 = [v_1, v_2, \ldots, v_k]^T$, $v^2 = [v_{k+1}, \ldots, v_m]^T$, $x^1 = [x_1, x_2, \ldots, x_k]^T$ and $x^2 = [x_{k+1}, \ldots, x_n]^T$. From (7.3.9) we have

$$x^1 = -A_{11}^{-1}v^1 - A_{11}^{-1}A_{12}x^2$$

and
$$v^2 + A_{21}x^1 + A_{22}x^2 = 0$$
which can be rearranged to give
$$\begin{bmatrix} x^1 \\ v^2 \end{bmatrix} + \begin{bmatrix} A_{11}^{-1} & A_{11}^{-1}A_{12} \\ -A_{21}A_{11}^{-1} & A_{22} - A_{21}A_{11}^{-1}A_{12} \end{bmatrix} \begin{bmatrix} v^1 \\ x^2 \end{bmatrix} = 0. \qquad (7.3.10)$$
The matrix in (7.3.10) is said to be obtained from A by *block pivoting* (A_{11} being the *pivotal submatrix*) and is combinatorially equivalent to A (see also [5]). When $k = 1$, (7.3.10) reduces to (7.3.5) with $i = j = 1$. By permuting rows and columns in (7.3.3) any k pairs of variables can be put into the desired positions in (7.3.9) and so can be interchanged provided the corresponding submatrix of A is nonsingular.

We can thus construct a family of totally unimodular matrices from any given one, and Balinski remarks [2] that the set of matrices which are combinatorially equivalent to P in Theorem 7.7 is identical with the most general known class of totally unimodular matrices [17]. However, examples of totally unimodular matrices which do not belong to this class can be constructed [22], and as in the case of unimodular vectors discussed in the previous section it seems likely that there is scope for further research.

EXERCISES

7.3.1 Use the definition provided by equation (7.3.8) to show that $B :: A$ if and only if $B = G^{-1}H$ where G is some $m \times m$ basis of $[I_m \ A]$ and H is the matrix formed by the remaining columns of $[I_m \ A]$.

7.3.2 Using equation (7.3.1) show that for any $m \times n$ matrix A, $-(A^{ij})^T = (-A^T)^{ji}$.

It then follows that $B :: A$ if and only if $-B^T :: -A^T$. Notice that this is not true if either the negative signs or transpose signs are omitted.

REFERENCES

1. BALINSKI, M. L., 'Integer programming: methods, uses, computation', *Management Sci.* **12**, 253–313 (1965).
2. BALINSKI, M. L., 'On finding integer solutions to linear programs', in *Proc. IBM Scientific Computing Symposium on combinatorial problems*, IBM, New York (1966).
3. BALINSKI, M. L. and SPIELBERG, K., 'Methods for integer programming: algebraic, combinatorial and enumerative', in *Progress in Operations Research*, vol. 3, Ed. J. S. Aronofsky, Wiley, New York (1969).
4. CAMION, P., 'Characterisation of totally unimodular matrices', *Proc. Amer. Math. Soc.* **16**, 1068–1073 (1965).

5. CEDERBAUM, I., 'Matrices all of whose elements and subdeterminants are 1, −1 or 0', *Jour. Math. Phys.* **36,** 351–361 (1958).
6. CEDERBAUM, I., 'Applications of matrix algebra to network theory', *IRE Trans. Circuit Theory,* **CT-6,** Special Supplement, 127–137 (1959).
7. CHANDRASEKARAN, R., 'Total unimodularity of matrices', *SIAM J. Appl. Math.* **17,** 1032–1034 (1969).
8. CHEEMA, M. S., 'Integral solutions of a system of linear equations', *Amer. Math. Monthly,* **73,** 487–490 (1966).
9. DANTZIG, G. B., *Linear programming and extensions,* Princeton University Press, New Jersey (1963).
10. FINKBEINER, D. T., *Introduction to matrices and linear transformations,* 2nd Edn., W. H. Freeman, San Francisco (1966).
11. GOMORY, R. E., 'On the relation between integer and noninteger solutions to linear programs', *Proc. Nat. Acad. Sci. USA,* **53,** 260–265 (1965).
12. GOMORY, R. E., 'Some polyhedra related to combinatorial problems', *Linear Algebra and its Applications,* **2,** 541–558 (1969).
13. HALEY, K. B., 'The solid transportation problem', *Opns. Res.* **10,** 448–463 (1962).
14. HALEY, K. B., 'The multi-index problem', *Opns. Res.* **11,** 368–379 (1963).
15. HELLER, I. and TOMPKINS, C. B., 'An extension of a theorem of Dantzig' in *Linear inequalities and related systems,* Ed. H. W. Kuhn and A. W. Tucker, Princeton University Press, New Jersey (1956).
16. HELLER, I., 'On linear systems with integral valued solutions', *Pacific J. Math.* **7,** 1351–1364 (1957).
17. HELLER, I. and HOFFMAN, A. J., 'On unimodular matrices', *Pacific J. Math.* **12,** 1321–1327 (1962).
18. HELLER, I., 'On unimodular sets of vectors' in *Recent advances in mathematical programming,* Ed. R. L. Graves and P. Wolfe, McGraw-Hill, New York (1963).
19. HELLER, I., 'On a class of equivalent systems of linear inequalities', *Pacific J. Math.* **13,** 1209–1227 (1963).
20. HELLER, I. 'On linear programs equivalent to the transportation program', *SIAM J. Appl. Math.* **12,** 31–42 (1964).
21. HOFFMAN, A. J. and KRUSKAL, J. B., 'Integral boundary points of convex polyhedra', in *Linear inequalities and related systems,* Ed. H. W. Kuhn and A. W. Tucker, Princeton University Press, New Jersey (1956).
22. HOFFMAN, A. J., 'Some recent applications of the theory of linear inequalities to extremal combinatorial analysis', in *Proc. symposia in applied math.* **10,** Combinatorial analysis, Ed.

R. Bellman and M. Hall, Jr., American Math. Society, Providence, Rhode Island (1960).
23. SMITH, H. J. S., 'On systems of linear indeterminate equations and congruences', *Philos. Trans. Roy. Soc. London*, **151,** 293–326 (1861).
24. TUCKER, A. W., 'A combinatorial equivalence of matrices', in *Proc. symposia in applied math.* **10,** Combinatorial analysis, Ed. R. Bellman and M. Hall, Jr., American Math. Society, Providence, Rhode Island (1960).
25. VEINOTT, A. F., Jr., and DANTZIG, G. B., 'Integral extreme points', *SIAM Review*, **10,** 371–372 (1968).

Appendix I

Notes on Algebra

A reasonable knowledge of matrix theory and basic algebra has been assumed throughout the book. We simply give here some results and definitions which are needed on occasion in the main body of the text, and which may not be at the reader's fingertips.

(1) If $A = [a_{ij}]$ is an $m \times n$ matrix and $B = [b_{ij}]$ an $l \times p$ matrix then the *Kronecker* (or *direct*) product of A and B, written $A \otimes B$, is the $ml \times np$ matrix partitioned as follows:

$$\begin{bmatrix} a_{11}B & a_{12}B & \dots & a_{1n}B \\ a_{21}B & a_{22}B & \dots & a_{2n}B \\ \cdot & \cdot & \dots & \cdot \\ a_{m1}B & a_{m2}B & \dots & a_{mn}B \end{bmatrix}.$$

If A is $n \times n$ and X is $n \times k$ then the equation $AX = C$ can be written $(A \otimes I_k)x = c$ where x is the column nk-vector composed of the rows of X taken in order, i.e.

$$x = [x_{11}, x_{12}, \dots, x_{1k}, x_{21}, x_{22}, \dots, x_{2k}, \dots, x_{n1}, \dots, x_{nk}]^T$$

and c is formed similarly from C.

The equation $YA = D$ can be written in a similar fashion as $(I_k \otimes A^T)y = d$, Y and D both being $k \times n$ matrices (see [1], [12]).

(2) A *permutation* of the integers 1 to n is an arrangement $\alpha_1, \alpha_2, \dots, \alpha_n$ in which each integer occurs exactly once. An $n \times n$ *permutation matrix* $P = [p_{ij}]$ is defined by $p_{ij} = 1$ when $j = \alpha_i$ and $p_{ij} = 0$ otherwise. In other words, the rows of P are rows $\alpha_1, \alpha_2, \dots, \alpha_n$ of I_n, so P has exactly one nonzero element in every row and column. P^T is also a permutation matrix and $P^TP = I_n$.

(3) A square matrix A is said to be *decomposable* if there exists a permutation matrix P such that

$$PAP^T = \begin{bmatrix} B & 0 \\ C & D \end{bmatrix}$$

where B and D are square. Otherwise A is *indecomposable.* Notice that PAP^T is the matrix obtained from A by replacing row i by row α_i, column i by column α_i $(i = 1, 2, \dots, n)$.

(4) A derogatory $n \times n$ matrix A is defined in Chapter 1 in terms of its Jordan form. An alternative definition which is sometimes

useful is as follows. Let the characteristic polynomial of A be

$$\det(\lambda I_n - A) = \lambda^n + a_1\lambda^{n-1} + \cdots + a_n \equiv p(\lambda).$$

The Cayley–Hamilton theorem states that

$$p(A) \equiv A^n + a_1 A^{n-1} + \cdots + a_n I_n = 0.$$

Let $b(\lambda) = \lambda^m + b_1\lambda^{m-1} + \cdots + b_m$ be the monic polynomial of least degree which annihilates A (i.e. $b(A) = 0$), called the *minimum polynomial* of A. If $m = n$, A is *nonderogatory*; if $m < n$ then A is *derogatory*. The minimum polynomial is always a factor of the characteristic polynomial, so a nonderogatory matrix has minimum and characteristic polynomials identical.

(5) An important example of a nonderogatory matrix is one which is in *companion form*, for example as in equation (2.2.1) (see Exercise 1.1.8). The name is also used for analogous matrices with the a_i in the first row, first column or last column (for example, the version of the matrix in (1.1.3) when $n = 1$). A given matrix is similar to a companion matrix associated with its characteristic polynomial if and only if it is nonderogatory.

(6) A *norm* is a real number which provides some measure of the 'size' of a matrix or vector. Many different definitions have been given but the only one used in this book is the *Euclidean norm* of an $m \times n$ matrix A,

$$\|A\| = \left(\sum_{i=1}^{m}\sum_{j=1}^{n}|a_{ij}|^2\right)^{1/2} = (\text{trace } A^*A)^{1/2}.$$

The name arises because the Euclidean norm of a vector v, $\|v\| = (\sum_i |v_i|^2)^{1/2}$, is what is normally understood as the length of v. For further material on norms, see [8].

(7) Let v^i $(i = 1, 2, \ldots, m)$ be a set of n-vectors and let

$$L = k_1 v^1 + k_2 v^2 + \cdots + k_m v^m$$

be a linear combination of these vectors, the k_i being drawn from the same number field as the components of the v^i. The vectors are *linearly independent* if $L = 0 \Rightarrow k_i = 0$ (all i) and *linearly dependent* otherwise. The rank of a matrix is equal to the greatest number of linearly independent rows or columns; this is equivalent to the definition that the rank is equal to the order of the largest nonzero minor.

(8) A set F of at least two distinct numbers is called a *number field* if whenever $a, b \in F$ then $a + b$, $a - b$, ab and a/b (provided $b \neq 0$) also $\in F$.

An expression

$$a(\lambda) = a_0\lambda^n + a_1\lambda^{n-1} + \cdots + a_n,$$

where the a_i belong to some number field F, is called a *polynomial over F*. The polynomial has *degree* δa equal to n, assuming $a_0 \neq 0$, and if $a_0 = 1$ is called *monic*. We have taken F to be the field of complex (or, in particular, real) numbers throughout this book. A *common divisor* of polynomials $a_1(\lambda), a_2(\lambda), \ldots, a_k(\lambda)$ is a polynomial which divides each $a_i(\lambda)$ (i.e., without remainder). The *greatest common divisor* (g.c.d.) is the unique monic common divisor of highest degree, and is itself divisible by any common divisor. If the g.c.d. has zero degree (i.e. is equal to unity) the polynomials are *relatively prime*.

The g.c.d. $d(\lambda)$ of two polynomials $a(\lambda)$ and $b(\lambda)$ can be expressed as

$$d(\lambda) = a(\lambda)x(\lambda) + b(\lambda)y(\lambda),$$

where $x(\lambda)$ and $y(\lambda)$ are also polynomials, in an infinite number of ways. However there is a unique pair having $\delta x < \delta b$, $\delta y < \delta a$, and when $d = 1$ this reduces to the result stated at the beginning of Section 2, Chapter 1. For further details see [2].

(9) The set Z of all integers and the set P of all polynomials with complex coefficients are both examples of a *principal ideal domain*. We have no space to go into any explanation of this concept and must refer the reader to a text on abstract algebra such as [11]. However, a fundamental property is that if a, b are any two nonzero members of a principal ideal domain D then their g.c.d. d can be written as $d = ax + by$, where x, y also belong to D (see (8) above). The fact that Z and P have this important property in common explains why integer matrices and polynomial matrices have the same type of canonical form (see Theorems 1.5 and 7.1).

(10) Some basic reference books on matrix theory are as follows: The outstanding work, generally regarded as the standard text in this field, is the translation of the two volumes by the Russian mathematician Gantmacher [7] who provides an extremely wide and detailed coverage; Mirsky [13] gives an excellent introduction which is clear and easy to read; the book by Bellman [1] reflects the invigorating attitude of its author and presents an original and stimulating approach with a variety of interesting applications; Ferrar's books [4, 5] are older but still often useful; other works worth noting are those by Cullen [3], Finkbeiner [6] and Lancaster [9], and a very well-written book by Noble [15]. Also of great value are the compendia of results due to MacDuffee [10] and more recently Marcus and Minc [12]; the survey [14] is more elementary but contains material on general algebra. Householder's book [8] on those parts of matrix theory of particular interest in numerical analysis is very useful, and for information on computation of characteristic roots and vectors Wilkinson's text [16] is indispensable.

REFERENCES

1. Bellman, R., *Introduction to matrix analysis*, 2nd Edn., McGraw-Hill, New York (1970).
2. Bôcher, M., *Introduction to higher algebra*, Dover, New York (1964) (reprint of 1907 Edn.).
3. Cullen, G. C., *Matrices and linear transformations*, Addison-Wesley, Reading, Massachusetts (1966).
4. Ferrar, W. L., *Algebra*, Oxford University Press (1941).
5. Ferrar, W. L., *Finite matrices*, Oxford University Press (1951).
6. Finkbeiner, D. T., *Introduction to matrices and linear transformations*, 2nd Edn., W. H. Freeman, San Francisco (1966).
7. Gantmacher, F. R., *The theory of matrices*, Vols. I and II, Chelsea Publishing Company, New York (1966).
8. Householder, A. S., *The theory of matrices in numerical analysis*, Blaisdell, Waltham, Massachusetts (1964).
9. Lancaster, P., *The theory of matrices*, Academic Press, New York (1969).
10. MacDuffee, C. C., *The theory of matrices*, Chelsea Publishing Company, New York (1956) (reprint of 1933 Edn.).
11. MacLane, S. and Birkhoff, G., *Algebra*, Macmillan, New York (1967).
12. Marcus, M. and Minc, H., *A survey of matrix theory and matrix inequalities*, Allyn and Bacon, Boston (1964).
13. Mirsky, L., *An introduction to linear algebra*, Oxford University Press (1963).
14. Mishina, A. P. and Proskuryakov, I. V., *Higher Algebra*, Pergamon Press, Oxford (1965).
15. Noble, B., *Applied linear algebra*, Prentice-Hall, Englewood Cliffs, New Jersey (1969).
16. Wilkinson, J. H., *The algebraic eigenvalue problem*, Oxford University Press (1965).

Appendix 2

Controllability and Observability of Linear Systems

The following ideas which we briefly introduce are referred to on a number of occasions throughout the book. Consider a system decribed by linear time-varying equations.

$$\left.\begin{aligned} \dot{x}(t) &= A(t)x(t) + B(t)u(t) \\ y(t) &= C(t)x(t) \end{aligned}\right\} \tag{A2.1}$$

where $A(t)$, $B(t)$, $C(t)$ are matrices having dimensions $n \times n$, $n \times l$ and $m \times n$ respectively. We say (A2.1) is *controllable* if, given any initial state $x(t_0) = x_0$, there exists a finite time $t_1 > t_0$ and a control $u(t)$ defined on $t_0 \leqslant t \leqslant t_1$ such that $x(t_1) = 0$. That is, it is possible to drive any state to the origin in a finite time. Sometimes controllability is taken to mean that the above criterion holds only for some x_0 and t_0, and *complete* controllability is used to refer to the situation which we have defined.

Let $\Phi(t, t_0)$ be the *transition* (or *fundamental*) matrix for the free system $\dot{x} = A(t)x$, so that

$$\dot{\phi}(t, t_0) = A(t)\Phi(t, t_0), \quad \Phi(t_0, t_0) = I_n.$$

Then Kalman [1] has proved

THEOREM A2.1 The system (A2.1) is controllable if and only if the symmetric matrix

$$\int_{t_0}^{t_1} \Phi(t_0, t)B(t)B^T(t)\phi^T(t_0, t)\, dt$$

is positive definite for some $t_1 > t_0$.

Notice that $\Phi(t_0, t) = \Phi^{-1}(t, t_0)$. In the case when A and B are independent of time, Theorem A2.1 reduces to

THEOREM A2.2 The constant linear system $\dot{x} = Ax + Bu$ (or the pair $[A, B]$) is controllable if and only if

$$\text{rank } [B, AB, A^2B, \ldots, A^{n-1}B] = n.$$

If rank $B = r$, this reduces to rank $[B, AB, \ldots, A^{n-r}B] = n$.

A number of alternative definitions of controllability have been suggested—see Chapter 5 of [3].

Observability means, roughly speaking, that the initial state of a system can be determined from a suitable measurement of the output $y(t)$. More precisely, the system (A2.1) is said to be *observable* if for any state $x(t_0) = x_0$ and given control vector $u(t)$, a knowledge of $y(t)$ on $t_0 \leqslant t \leqslant t_1$ is sufficient to determine x_0.

The system

$$\left.\begin{aligned} \dot{\tilde{x}}(t) &= A^T(t)\tilde{x}(t) + C^T(t)\tilde{u}(t) \\ \tilde{y}(t) &= B^T(t)\tilde{x}(t), \end{aligned}\right\} \qquad \text{(A2.2)}$$

where $\tilde{u}(t)$ is a column m-vector and $\tilde{y}(t)$ a column l-vector, is called the *dual* of (A2.1), and we have the result:

THEOREM A2.3 The system (A2.1) is observable if and only if its dual is controllable.

Thus for constant linear systems:

THEOREM A2.4 The pair $[A, C]$ is observable if and only if $[A^T, C^T]$ is controllable; i.e. if and only if

$$\text{rank}\,[C^T, A^TC^T, \ldots, (A^T)^{n-1}C^T] = n.$$

Again, other definitions have been given, but the preceding results are sufficient for the purposes of this book. The references listed should be consulted for further details.

REFERENCES

1. KALMAN, R. E., 'Contributions to the theory of optimal control', *Bol. Soc. Mat. Mex.* **5,** 102–119 (1960).
2. KREINDLER, E. and SARACHIK, P. E., 'On the concepts of controllability and observability of linear systems', *IEEE Trans. Aut. Control,* **AC-9,** 129–136 (1964).
3. ROSENBROCK, H. H., *State-space and multivariable theory,* Nelson, London (1970).
4. SILVERMAN, L. M. and MEADOWS, H. E., 'Controllability and observability in time-variable linear systems', *SIAM J. Control,* **5,** 64–73 (1967).
5. ZADEH, L. A. and DESOER, C. A., *Linear system theory,* McGraw-Hill, New York (1963).

Appendix 3
Stability Definitions and Theorems

We present a short introduction to some concepts of stability for the set of ordinary differential equations

$$\dot{x}_i = f_i(x_1, x_2, \ldots, x_n), \qquad (i = 1, 2, \ldots, n), \qquad \text{(A3.1)}$$

useful as background material for Section 4.1. We assume that the equations have a solution $x(t) = [x_1(t), \ldots, x_n(t)]$ satisfying the initial condition $x(0) = x_0$, and that there is an *equilibrium* (or *critical*) *point* at the origin, i.e. $f(0) = 0$, $t \geqslant 0$. This equilibrium point is said to be *stable* if, given any region R_1 around the origin, another region R_2 can be found such that any solution starting inside R_2 never leaves R_1. The origin is said to be *asymptotically stable* if it is stable and if in addition $x(t) \longrightarrow 0$ as $t \longrightarrow \infty$. Roughly speaking, stability implies that after perturbation of the system finite oscillations may occur about the equilibrium point, but asymptotic stability ensures that these will eventually decay to zero. The practical usefulness of these particular definitions, however, is limited by the fact that R_2 (i.e. the region of permissible perturbations from equilibrium) may be very small.

Liapunov's 'second' or 'direct' method is a means of determining stability of (A3.1) without actually solving the equations. It relies on the concept of a *positive* (*negative*) *definite function* of x which is a scalar function taking positive (negative) values at all points except the origin, where it is zero. A *semidefinite* function similarly does not change sign but may be zero at points other than the origin. More generally, a function $h(x, t)$ is said to be positive definite if $h(0, t) = 0$ and if there exists a positive definite function $k(x)$ such that $h(x, t) \geqslant k(x)$ in some region $\|x\| \leqslant d$, $t \geqslant 0$. In particular a symmetric matrix $K(t)$ is called positive definite if the time-varying quadratic form $x^T(t)K(t)x(t)$ is positive definite (see Theorem 5.4). Liapunov's basic result states simply:

THEOREM A3.1 If a positive definite function $V(x)$ can be found whose time derivative with respect to (A3.1), namely

$$\dot{V} = \sum_{i=1}^{n} (\partial V/\partial x_i)\dot{x}_i = \sum_{i=1}^{n} (\partial V/\partial x_i)f_i$$

is negative semidefinite, then the origin of (A3.1) is stable. If $\dot{V}$ is negative definite the origin is asymptotically stable.

In these circumstances $V(x)$ is called a *Liapunov function*. For the system of constant linear equations (4.1.1), choosing a quadratic form $V(x) = x^T Px$ as a possible Liapunov function gives

$$\dot{V} = \dot{x}^T Px + x^T P\dot{x} = x^T(A^T P + PA)x$$

and setting $\dot{V} = -x^T Qx$ leads immediately to Theorem 4.2. Remark (iii) after Theorem 4.2 relies on a theorem of Liapunov on instability, which requires that $\dot{V}$ must be negative definite and that there must be a region around the origin in which V is also negative.

Precise mathematical definitions of the preceding and other types of stability, and rigorous statements, proofs and generalizations of the theorems can be found in the textbooks listed below.

REFERENCES

1. HAHN, W., *Theory and application of Liapunov's direct method*, Prentice-Hall, Englewood Cliffs, New Jersey (1963).
2. ZUBOV, V. I., *Methods of A. M. Liapunov and their application*, Noordhoff, Groningen (1964).

Appendix 4

Optimal Control Theory

We give here a very brief and simplified summary of some of the background theory necessary for the derivation of the Riccati equation in Section 5.2. For further details we refer the reader to [1].

Consider a control system described by the differential equations

$$\dot{x} = f(x, u) \tag{A4.1}$$

where $x = [x_1, \ldots, x_n]^T, f = [f_1, \ldots, f_n]^T$ and $u = [u_1, \ldots, u_m]^T$, x and u being respectively the state and control vectors. It is desired to choose the components of u so that some measure of the system performance, which we shall take as

$$J = K[x(t_1)] + \int_{t_0}^{t_1} L[x(t), u(t)]\, dt, \tag{A4.2}$$

where K and L are scalar functions, is minimized. We shall assume that t_0, t_1 and $x(t_0)$ are specified but that the final state $x(t_1)$ is unknown. Of course, this is only a particular example of a wide variety of possible end conditions. The Pontryagin method (which is a generalization of the classical calculus of variations) for solving this problem is as follows. Introduce a function, termed the *Hamiltonian*,

$$H = L(x, u) + p^T f, \tag{A4.3}$$

and a set of associated differential equations

$$\left.\begin{aligned} \dot{x}_i &= \frac{\partial H}{\partial p_i} & \text{(A4.4a)} \\ \dot{p}_i &= -\frac{\partial H}{\partial x_i} & \text{(A4.4b)} \end{aligned}\right\} i = 1, 2, \ldots, n.$$

The vector $p = [p_1, \ldots, p_n]^T$ is called the *adjoint* (or costate) vector, and (A4.4b) the *adjoint* equations. Notice that (A4.4a) simply reduces to the original equations (A4.1) on using (A4.3). The solution to (A4.4) is required to satisfy the two-point boundary conditions

$$x(t_0) = x_0, p_i(t_1) = \left[\frac{\partial K}{\partial x_i}\right]_{t=t_1}, \quad (i = 1, 2, \ldots, n). \tag{A4.5}$$

Pontryagin's *maximum principle* states that (A4.2) will be minimized if u is chosen so as to maximize H.

We now apply this to the linear control system with quadratic performance index considered in Section 5.2, namely

$$\dot{x}(t) = A(t)x(t) + B(t)u(t)$$
$$x(t_0) = x_0 \tag{A4.6}$$

$$\mathcal{J} = \tfrac{1}{2}x^T(t_1)Mx(t_1) + \tfrac{1}{2}\int_{t_0}^{t_1} [x^T(t)Q(t)x(t) + u^T(t)R(t)u(t)]\, dt. \tag{A4.7}$$

The Hamiltonian in (A4.3) is

$$H = \tfrac{1}{2}x^TQx + \tfrac{1}{2}u^TRu + p^TAx + p^TBu \tag{A4.8}$$

and (A4.4b) gives

$$\dot{p} = -Qx - A^Tp. \tag{A4.9}$$

Setting the derivative of H in (A4.8) with respect to u equal to zero in order to find an extreme value, we obtain $Ru + B^Tp = 0$, so that

$$u = -R^{-1}B^Tp. \tag{A4.10}$$

Equations (A4.6) and (A4.9) then become

$$\left.\begin{aligned} \dot{x} &= Ax - BR^{-1}B^Tp \\ \dot{p} &= -Qx - A^Tp \end{aligned}\right\} \tag{A4.11}$$

with boundary conditions given by (A4.5) as

$$x(t_0) = x_0, p(t_1) = Mx(t_1). \tag{A4.12}$$

It can be shown that the unique optimal control is obtained by setting $p(t) = P(t)x(t)$, where $P(t)$ is a symmetric matrix. Substitution into (A4.11) then leads to

$$\dot{P}x + P(Ax - BR^{-1}B^TPx) = -Qx - A^TPx,$$

which must hold for all $x(t)$, showing that $P(t)$ satisfies

$$\dot{P} = PBR^{-1}B^TP - A^TP - PA - Q \tag{A4.13}$$

with boundary condition obtained from (A4.12) as $P(t_1) = M$. Thus the quadratic performance index (A4.7) leads to linear feedback

$$u = -R^{-1}(t)B^T(t)P(t)x(t)$$

where $P(t)$ is the unique positive definite solution of the Riccati equation (A4.13) satisfying the stated condition. The minimum value of (A4.7) is then $\frac{1}{2}x_0^TP(t_0)x_0$.

REFERENCE

1. ATHANS, M. and FALB, P. L., *Optimal Control*, McGraw-Hill, New York (1966).

Appendix 5
Additional Bibliography

We give here a few selected references which for one reason or another were not included in the main body of the text but which are nevertheless worth consulting.

CHAPTER 1

GUDERLEY, K. G., 'On nonlinear eigenvalue problems for matrices', *SIAM J. Appl. Math.* **6**, 335–353 (1958).

HENNION, P. E., 'Reduction of a matrix containing polynomial elements', *Comm. ACM*, **6**, 165–166 (1963).

LANCASTER, P. and WEBBER, P. N., 'Jordan chains for lambda-matrices', *Linear Algebra and its Applications*, **1**, 563–569 (1968).

CHAPTER 2

CUTTERIDGE, O. P. D., 'Some tests for the number of positive zeros and for the number of real and complex zeros of a real polynomial', *Proc. IEE*, **107C**, 105–110 (1960).

HOUSEHOLDER, A. S., *The numerical treatment of a single non-linear equation*, McGraw-Hill, New York (1970).

KU, S. Y. and ADLER, R. J., 'Computing polynomial resultants: Bezout's determinant vs. Collins' reduced P.R.S. algorithm', *Comm. ACM*, **12**, 23–30 (1969).

LEPSCHY, A., 'A new criterion for evaluating the number of complex roots of an algebraic equation', *Proc. IEEE*, **50**, 1981 (1962).

CHAPTER 3

ANDERSON, B. D. O., 'The testing for optimality of linear systems', *Int. J. Control*, **4**, 29–40 (1966).

ANDERSON, B. D. O., 'Algebraic description of bounded real matrixes', *Electronics Letters*, **2**, 464 (1966).

ANDERSON, B. D. O., 'The inverse problem of stationary covariance generation', *J. Statistical Physics*, **1**, 133–147 (1969).

DOWNS, T., 'On the inversion of a matrix of rational functions', *Linear Algebra and its Applications*, **4**, 1–10 (1971).

GUEGUEN, C. J. and TOUMIRE, E., 'Comments on: Irreducible Jordan form realization of a rational matrix', *IEEE Trans. Aut. Control*, **AC-14**, 783–784 (1969).

HEYMANN, M., 'A unique canonical form for multivariable linear systems', *Int. J. Control*, **12**, 913–927 (1970).

NEWCOMB, R. W. and ANDERSON, B. D. O., 'On the generation of all spectral factors', *IEEE Trans. Information Theory*, **IT-14**, 512–514 (1968).

WOLOVICH, W. A. and FALB, P. L., 'On the structure of multivariable systems', *SIAM J. Control*, **7**, 437–451 (1969).

YOULA, D. C., 'The synthesis of linear dynamical systems from prescribed weighing patterns', *SIAM J. Appl. Math.* **14**, 527–549 (1966).

CHAPTER 4

BERMAN, A. and BEN-ISRAEL, A., 'Linear equations over cones with interior: a solvability theorem with applications to matrix theory', *Rpt. No. 69-1, Series in Applied Math.*, Northwestern University, Evanston, Illinois (1969).

BERMAN, A. and BEN-ISRAEL, A., 'More on linear inequalities with applications to matrix theory', *J. Math. Anal. Appl.* 33, 482–496 (1971).

MULLER, P. CHR., 'Solution of the matrix equations $AX + XB = -Q$ and $S^TX + XS = -Q$', *SIAM J. Appl. Math.* **18**, 682–687 (1970).

SCHNEIDER, H., 'Positive operators and an inertia theorem', *Num. Math.* **7**, 11–17 (1965).

SNYDERS, J. and ZAKAI, M., 'On nonnegative solutions of the equation $AD + DA' = -C$', *SIAM J. Appl. Math.* **18**, 704–714 (1970).

TAUSSKY, O., 'Positive definite matrices and their role in the study of characteristic roots of general matrices', *Advances in Math.* **2**, 175–186 (1968).

CHAPTER 5

FALB, P. L. and KLEINMAN, D. L., 'Remarks on the infinite dimensional Riccati equation', *IEEE Trans. Aut. Control*, **AC-11**, 534–536 (1966).

KREINDLER, E. and HEDRICK, J. K., 'On equivalence of quadratic loss functions', *Int. J. Control*, **11**, 213–222 (1970).

MAN, F. T., 'Comments on: Solution of the algebraic matrix Riccati equation via Newton–Raphson iteration', *AIAA J.*, **6,** 2463–2464 (1968).

MOORE, J. B., 'Application of Riccati equations in systems engineering', *Instn. of Engineers, Australia, Elec. Eng. Trans.* **EE5,** 29–34 (1969).

REID, W. T., 'Properties of solutions of a Riccati matrix differential equation', *J. Math. and Mech.* **9,** 749–770 (1960).

WALTER, O. H. D., 'Formulas for linear optimal control over a finite or infinite interval', *Electronics Letters*, **6,** 632–633 (1970).

WONHAM, W. M., 'On a matrix Riccati equation of stochastic control', *SIAM J. Control*, **6,** 681–697 (1968).

CHAPTER 6

ALBERT, A., 'Conditions for positive and nonnegative definiteness in terms of pseudoinverses', *SIAM J. Appl. Math.* **17,** 434–440 (1969).

ANDERSON, W. N. Jr. and DUFFIN, R. J., 'Series and parallel addition of matrices', *J. Math. Anal. Appl.* **26,** 576–594 (1969).

BEN-ISRAEL, A., 'On error bounds for generalized inverses', *SIAM J. Numer. Anal.* **3,** 585–592 (1966).

BOULLION, T. L. and ODELL, P. L. (Eds.), *Proceedings of the symposium on theory and application of generalized inverses of matrices*, Texas Technological College Press, Lubbock, Texas (1968).

CHARNES, A. and KIRBY, M., 'Modular design, generalized inverses and convex programming', *Opns. Res.* **13,** 836–847 (1965).

HURT, M. F. and WAID, C., 'A generalized inverse which gives all the integral solutions to a system of linear equations', *SIAM J. Appl. Math.*, **19,** 547–550 (1970),

KATZ, I. J. and PEARL, M. H., 'On *EPr* and normal *EPr* matrices', *J. Res. NBS*, **70B,** 47–77 (1966).

MEYER, C. D., 'Generalized inverses of triangular matrices', *SIAM J. Appl. Math.* **18,** 401–406 (1970).

CHAPTER 7

CAMION, P., 'Matrices totalement unimodulaires et problèmes combinatoires', *Eur. 1632f*, Centre Commun de Recherche Nucléaire (1964).

SAATY, T. L., *Optimization in integers and related extremal problems*, McGraw-Hill, New York (1970).

VEBLEN, O. and FRANKLIN, P., 'On matrices whose elements are integers', *Ann. Math.* **23,** 1–15 (1921–22).

Solutions to Exercises

The value of these solutions will of course be greatly diminished if they are consulted before a serious attempt has been made at the problems. Please do not cheat!

CHAPTER 1

1.1.1 For division on the right, let

$$B(\lambda) = B_0\lambda^k + B_1\lambda^{k-1} + \cdots + B_k$$
$$= (Q_0\lambda^{k-1} + Q_1\lambda^{k-2} + \cdots + Q_{k-1})(\lambda I - A) + R.$$

Equating powers of λ gives

$$B_0 = Q_0,\ B_1 = Q_1 - Q_0A,\ B_2 = Q_2 - Q_1A, \ldots,$$
$$B_k = R - Q_{k-1}A.$$
$$\therefore\ Q_1 = B_1 + B_0A,\ Q_2 = B_2 + B_1A + B_0A^2, \ldots,$$
$$Q_{k-1} = B_{k-1} + B_{k-2}A + \cdots + B_0A^{k-1},$$

hence $R = B_k + B_{k-1}A + \cdots + B_0A^k$.
Similarly for division on the left.

1.1.2 $$R(\lambda)T(\lambda) = (I + \lambda R_1)(T_0\lambda^{n-1} + T_1\lambda^{n-2} + \cdots + T_{n-2}\lambda + I)$$
$$= R_1T_0\lambda^n + (T_0 + R_1T_1)\lambda^{n-1} + (T_1 + R_1T_2)\lambda^{n-2} + \cdots + (R_1T_{n-2} + T_{n-3})\lambda^2 + (T_{n-2} + R_1)\lambda + I.$$

Set $T_{n-2} = -R_1,\ T_{n-3} = -R_1T_{n-2} = R_1^2, \ldots,$
$$T_1 = (-1)^{n-2}R_1^{n-2},$$
$$T_0 = (-1)^{n-1}R_1^{n-1}.$$

Hence $R_1T_0 = (-1)^{n-1}R_1^n = 0$, so $R(\lambda)T(\lambda) = I$.

1.1.3 If X_1, X_2 are invertible, $\det X_1X_2 = \det X_1 \det X_2 =$ constant, so X_1X_2 is also invertible.
Hence, with

$$R(\lambda) = I + R_1\lambda,\ S(\lambda) = I + S_1\lambda \text{ and } R_1^n = S_1^n = 0,$$

then

$$R^2 = R_1^2\lambda^2 + 2R_1\lambda + I \text{ and } RS = R_1S_1\lambda^2 + (R_1 + S_1)\lambda + I$$

are both invertible.

1.1.4 $R(\lambda)T(\lambda) = I$, where $T(\lambda)$ is also a polynomial matrix, and differentiating:

$$(dR/d\lambda)T + R(dT/d\lambda) = 0$$
$$\therefore\ (dR/d\lambda) = -R(dT/d\lambda)T^{-1} = -R(dT/d\lambda)R.$$

$\therefore$ by Theorem 1.3, $dR/d\lambda$ and $dT/d\lambda$ are equivalent.

1.1.5 From

$$(N_0\lambda^p + N_1\lambda^{p-1} + \cdots + N_p)(\nu_0\lambda^q + \nu_1\lambda^{q-1} + \cdots + \nu_q) = I$$

we have

$N_0\nu_0 = 0$, $N_1\nu_0 + N_0\nu_1 = 0$ (coefficients of λ^{p+q}, λ^{p+l-1}).

Therefore

$$\begin{aligned}(N_0\lambda + N_1)(\nu_0\lambda + \nu_1) &= N_0\nu_0\lambda^2 + (N_1\nu_0 + N_0\nu_1)\lambda + N_1\nu_1 \\ &= N_1\nu_1,\end{aligned}$$

so $N_0\lambda + N_1$ and $\nu_0\lambda + \nu_1$ will be invertible provided N_1, ν_1 are nonsingular.

1.1.6

$$\begin{aligned} &\begin{vmatrix} 0 & -I & I\lambda \\ -I & I\lambda - A_1 & A_1\lambda \\ I\lambda & A_1\lambda & A_2\lambda + A_3 \end{vmatrix} \\ &= \begin{vmatrix} 0 & -I & I\lambda \\ -I & I\lambda - A_1 & A_1\lambda \\ 0 & I\lambda^2 & A_1\lambda^2 + A_2\lambda + A_3 \end{vmatrix} \\ &= \begin{vmatrix} -I & I\lambda \\ I\lambda^2 & A_1\lambda^2 + A_2\lambda + A_3 \end{vmatrix} \quad \text{expanding by 1st block column} \\ &= | -A_1\lambda^2 - A_2\lambda - A_3 - I\lambda^3 | \\ &= (-1)^n \, | I\lambda^3 + A_1\lambda^2 + A_2\lambda + A_3 |. \end{aligned}$$

1.1.7 Setting $\mu = 1/\lambda$ in equation (1.1.1) ($\lambda = 0$ is not a latent root since $|A_N| \neq 0$) gives

$$(I\mu^N + A_{N-1}A_N^{-1}\mu^{N-1} + \cdots + A_0A_N^{-1})A_N/\mu^N.$$

Application of Theorem 1.7 to the matrix within the brackets gives $\det(\mu I_{nN} - D) = \det(I\mu^N + \cdots + A_0A_N^{-1})$ where D is the $nN \times nN$ matrix as in equation (1.1.3) but with last column $-A_0A_N^{-1}, \ldots, -A_{N-1}A_N^{-1}$.

Therefore

$$\det(I_{nN}/\lambda - D) = \det(A(\lambda)A_N^{-1}/\lambda^N)$$

or

$$[\det(I_{nN} - \lambda D)]/\lambda^{nN} = \det A(\lambda)/(\det A_N)\lambda^{Nn}$$

$\therefore$ $\det A(\lambda) = \det A_N \det(I_{nN} - \lambda D)$.

1.1.8 Since $n = 1$, C is the $N \times N$ matrix

$$\begin{bmatrix} 0 & 0 & & & -a_N \\ 1 & 0 & & & -a_{N-1} \\ 0 & 1 & & & -a_{N-2} \\ & & \cdot & & \cdot \\ & & & \cdot & \cdot \\ & & & \cdot & \cdot \\ & & & 1 & -a_1 \end{bmatrix}$$

$$\text{and } \lambda I_N - C = \begin{bmatrix} \lambda & 0 & & & & a_N \\ -1 & \lambda & & & & a_{N-1} \\ 0 & -1 & & & & a_{N-2} \\ & & \cdot & & & \cdot \\ & & & \cdot & & \cdot \\ & & & & \cdot & \cdot \\ & & & & -1 & \lambda + a_1 \end{bmatrix}.$$

Add λ times second row onto first, then interchange first and second rows to get

$$\begin{matrix} -1 & \lambda & 0 & \ldots & a_{N-1} \\ 0 & \lambda^2 & 0 & \ldots & a_N + \lambda a_{N-1} \end{matrix}$$

as the new first two rows. Add λ times the first column to the second column, a_{N-1} times the first column to the last column to get the following matrix, equivalent to $\lambda I_N - C$:

$$\begin{bmatrix} -1 & 0 & 0 & . & . & . & 0 \\ 0 & \lambda^2 & 0 & . & . & . & a_N + \lambda a_{N-1} \\ 0 & -1 & \lambda & & & & a_{N-2} \\ & & -1 & & & & \cdot \\ & & & \cdot & & & \cdot \\ & & & & \cdot & & \cdot \\ & & & & & \cdot & \lambda + a_1 \end{bmatrix}$$

Continue by adding $\lambda^2 \times$ row 3 onto row 2 and interchanging rows 2 and 3, etc., to end up with diag $[1, 1, \ldots, 1, a_N + \lambda a_{N-1} + \cdots + \lambda^N]$.

1.2.1 If $X(\lambda)$ and $Y(\lambda)$ are matrices satisfying $AX + YB = E$, then

$$\begin{bmatrix} I & -Y \\ 0 & I \end{bmatrix}\begin{bmatrix} A & E \\ 0 & B \end{bmatrix}\begin{bmatrix} I & -X \\ 0 & I \end{bmatrix} = \begin{bmatrix} A & E - AX - YB \\ 0 & B \end{bmatrix}$$
$$= \begin{bmatrix} A & 0 \\ 0 & B \end{bmatrix},$$

showing equivalence.

1.2.2 Since the characteristic roots of $\tilde{A} = A^2 + b_1 A + b_2 I$ are $\lambda_j^2 + b_1\lambda_j + b_2$, where λ_j are the roots of A, if $\tilde{A}$ is singular there must be some λ_j for which $\lambda_j^2 + b_1\lambda_j + b_2 = 0$. Hence Re $(\lambda_j) < 0$ since $b_1 > 0$, $b_2 > 0$.

Conversely, if some $\lambda_j = -\alpha + i\beta$ $(\alpha > 0)$, then a characteristic root of $\tilde{A}$ is

$$(-\alpha + i\beta)^2 + b_1(-\alpha + i\beta) + b_2$$
$$= \alpha^2 - \beta^2 - b_1\alpha + b_2 + i(-2\alpha\beta + b_1\beta).$$

Taking $\quad b_1 = 2\alpha, b_2 = \alpha^2 + \beta^2 \ (\beta \neq 0)$

or $\quad b_1 > \alpha, \ b_2 = b_1\alpha - \alpha^2 \ (\beta = 0),$

gives $\tilde{A}$ a zero characteristic root.

1.3.1 When $n = m = 1$ in Theorem 1.13, the two scalar polynomials are

$$a(\lambda) = \lambda^N + a_1\lambda^{N-1} + \cdots + a_N$$

and
$$b(\lambda) = \lambda^M + b_1\lambda^{M-1} + \cdots + b_M.$$

The $N \times N$ matrix C is the companion matrix of $a(\lambda)$, given in the solution to Exercise 1.1.8. The matrix R in equation (1.3.1) is $C^M + b_1C^{M-1} + \cdots + b_MI_N$, and Theorem 1.13 reduces to the result that the degree of the g.c.d. of $a(\lambda)$ and $b(\lambda)$ is $N -$ rank R.

1.3.2 Apply Theorem 1.13 with $M = N = 1$, $m = n$ and $C = A$ so that $R = I_n \otimes A - B \otimes I_n$, giving the desired result.

1.4.1
$$\begin{bmatrix} T_2 & U_2 \\ -V_2 & W_2 \end{bmatrix} = \begin{bmatrix} M_1 & 0 \\ N_1 & I_m \end{bmatrix}\begin{bmatrix} T_1 & U_1 \\ -V_1 & W_1 \end{bmatrix}\begin{bmatrix} M_2 & N_2 \\ 0 & I_l \end{bmatrix}$$
$$= \begin{bmatrix} M_1T_1M_2 & M_1T_1N_2 + M_1U_1 \\ (N_1T_1 - V_1)M_2 & (N_1T_1 - V_1)N_2 + N_1U_1 + W_1 \end{bmatrix}.$$

Hence the transfer function matrix for $P_2(\lambda)$ is

$$\begin{aligned} G_2 &= V_2T_2^{-1}U_2 + W_2 \\ &= (V_1 - N_1T_1)M_2M_2^{-1}T_1^{-1}M_1^{-1}(M_1T_1N_2 + M_1U_1) \\ &\qquad + (N_1T_1 - V_1)N_2 + N_1U_1 + W_1 \\ &= V_1T_1^{-1}U_1 + W_1 = G_1. \end{aligned}$$

1.4.2 Inserting the given expressions for $x(\lambda)$, $y(\lambda)$ and $z(\lambda)$ into (1.4.12), and equating powers of λ, gives

$$\begin{aligned} x_{n-2} \quad &+ By_{n-1} &&= z_{n-1} \\ -Ax_{n-2} + x_{n-3} \quad &+ By_{n-2} &&= z_{n-2} \\ &\vdots && \\ -Ax_1 + x_0 \quad &+ By_1 &&= z_1 \\ -Ax_0 \quad &+ By_0 &&= z_0. \end{aligned} \tag{i}$$

If $\lambda I_n - A$ and B are relatively prime, this set of equations does have a solution, because of the non-vanishing of the determinant of coefficients (Theorem 1.21).

If $(\lambda I - A)x^{(i)} + By^{(i)} = e_i$, then

$$(\lambda I - A)[x^{(1)}, x^{(2)}, \ldots, x^{(n)}] + B[y^{(1)}, y^{(2)}, \ldots, y^{(n)}] = [e_1, e_2, \ldots, e_n].$$

1.4.3 If (1.4.13) has a solution, then the equations (i) in the preceding solution hold with the right-hand side replaced by $0, 0, \ldots, 0, e_i$. Add A times the first equation to the second, then A times the second to the third, etc., to obtain finally

$$A^{n-1}By_{n-1} + A^{n-2}By_{n-2} + \cdots + By_0 = e_i,$$

or

$$[A^{n-1}B, A^{n-2}B, \ldots, B]\begin{bmatrix} y_{n-1} \\ \cdot \\ \cdot \\ \cdot \\ y_0 \end{bmatrix} = e_i.$$

Since this holds for each $i = 1, 2, \ldots, n$ (of course with different y_j) it follows that rank $[A^{n-1}B, A^{n-2}B, \ldots, B] = n$, so by Theorem 1.22, $\lambda I_n - A$ and B are relatively prime.

1.4.4 From Theorem 1.18, since $B(\lambda)$ and $\lambda I_n - A$ are relatively prime there exist $X(\lambda)$, $Y(\lambda)$ satisfying

$$BX + (\lambda I - A)Y = I.$$

If $A^kB_0 + A^{k-1}B_1 + \cdots + B_k = 0$, then from Exercise 1.1.1, $B = (\lambda I - A)Q(\lambda)$ which implies

$$(\lambda I - A)QX + (\lambda I - A)Y = I,$$

or

$$(\lambda I - A)(QX + Y) = I,$$

showing $\lambda I - A$ is invertible, a contradiction.

1.4.5 Let $U_0[T \quad U]\begin{bmatrix} V_1 & V_2 \\ V_3 & V_4 \end{bmatrix} = [S \quad 0]$, the Smith form of $[T \quad U]$.

Multiplying out gives

$$U_0(TV_1 + UV_3) = S$$

or

$$TV_1 + UV_3 = U_0^{-1}S$$
$$= D, \text{ another polynomial matrix since } U_0 \text{ is invertible.}$$

$\therefore$ If $T = D_iT_i$, $U = D_iU_i$ where $D_i(\lambda)$ is any left divisor of T and U, then

$$D_i(T_iV_1 + U_iV_3) = D. \qquad \text{(ii)}$$

Also, since $\begin{bmatrix} V_1 & V_2 \\ V_3 & V_4 \end{bmatrix}$ is invertible, we have

$$[T \quad U] = U_0^{-1}[S \quad 0]\begin{bmatrix} Z_1 & Z_2 \\ Z_3 & Z_4 \end{bmatrix}$$

$\therefore$

$$T = U_0^{-1}SZ_1 = DZ_1$$

and

$$U = U_0^{-1}SZ_2 = DZ_2.$$

$\therefore$ D is a common left divisor of T and U, and must be a greatest common left divisor because of (ii).

1.4.6 $G_1 = VT_1^{-1}U_1 + W$, $G_2 = V_2T_2^{-1}U + W$, and $T_1 = D^{-1}T$, $U_1 = D^{-1}U$, $T_2 = TE^{-1}$, $V_2 = VE^{-1}$.
Substitution gives $G_1 = VT^{-1}U + W = G_2$, as required.

For first system,

$$\text{order} = \delta(\det T_1) = \delta(\det T) - \delta(\det D) < \delta(\det T)$$

provided $\delta(\det D) > 0$.

Similarly for $\delta(\det T_2)$.

1.5.1 Using equation (1.5.15a),

$$B = \begin{bmatrix} 3 & -1 & 0 \\ -2 & 4 & 0 \\ -4 & 3 & 1 \end{bmatrix} \begin{bmatrix} 2 & -1 & 0 \\ 4 & -2 & 0 \\ 6 & -3 & 0 \end{bmatrix}$$

$$= \begin{bmatrix} 2 & -1 & 0 \\ 12 & -6 & 0 \\ 10 & -5 & 0 \end{bmatrix}.$$

$[\theta_2, \theta_3, \theta_6]^T = A^{-1}Bx^0 = [3, 6, 9]^T$, so from equations (1.5.9) and (1.5.10), $\bar{\theta} = \min(4/3, 5/6, 11/9) = 5/6$ and $\underline{\theta} = -\infty$.

Also $[\mu_1, \mu_4, \mu_5] = c_x A^{-1}BA^{-1}N$

$$= [1, -3, 0]A^{-1}BA^{-1}\begin{bmatrix} 1 & 0 & 2 \\ 0 & 1 & 0 \\ 0 & 0 & 8 \end{bmatrix}$$

$$= [-3, 1/2, -6],$$

and equations (1.5.11) and (1.5.12) give

$$\bar{\mu} = \min(1/15, 2/5) = 1/15, \ \underline{\mu} = -8/5.$$

Hence from (1.5.8),

$$-8/5 \leqslant \lambda \leqslant 1/15.$$

The solution to the parametric problem is $x^0 - \lambda\theta$, so $x_2 = 4 - 3\lambda$, $x_3 = 5 - 6\lambda$, $x_6 = 11 - 9\lambda$ and the corresponding minimum value of z is $z^0 - \lambda c_x\theta = -11 + 15\lambda$.

1.5.2 When A is fixed, N changes to $N + \lambda M$, then $y = 0$ gives $x = x^0$ as a feasible solution for any value of λ. Equation (1.5.20) is replaced by

$$z = c_x A^{-1}B + (c_y - c_x A^{-1}N - \lambda c_x A^{-1}M)y$$

$$= c_x A^{-1}B + \sum_{i=m+1}^{n} (\gamma_i - \lambda\delta_i)x_i,$$

so λ is determined by $\gamma_i - \lambda\delta_i \geqslant 0$, $i = m + 1, \ldots, n$.

1.5.3 Equation (1.5.17) becomes

$$x = A^{-1}(I - \lambda BA^{-1})b - A^{-1}(I - \lambda BA^{-1})(N + \lambda M)y, \qquad \text{(iii)}$$

where B satisfies (1.5.15a) or (1.5.15b). Therefore, for no second degree term, $BA^{-1}M = 0$. The θ_i in (1.5.18) are unaltered since we set $y = 0$ in (iii). Also,

$$x = A^{-1}(I - \lambda BA^{-1})b - (A^{-1}N - \lambda P)y,$$

so the second term in (1.5.20) becomes $[c_y - c_x(A^{-1}N - \lambda P)]y$.

CHAPTER 2

2.1.1 Using (2.1.7), (i)

$$D = \frac{1}{a}\begin{vmatrix} a & b & c \\ 0 & 2a & b \\ 2a & b & 0 \end{vmatrix}$$
$$= b^2 - 4ac.$$

(ii)

$$D = \begin{vmatrix} 1 & 0 & 3b & c & 0 \\ 0 & 1 & 0 & 3b & c \\ 0 & 0 & 3 & 0 & 3b \\ 0 & 3 & 0 & 3b & 0 \\ 3 & 0 & 3b & 0 & 0 \end{vmatrix}$$
$$= -27(c^2 + 4b^3).$$

2.1.2 Result follows by expanding determinant in (2.1.11) by last column, since all cofactors are independent of λ.

2.1.3 It is well known, and easy to show, that

$$\det V = \prod_{i<j} (\alpha_j - \alpha_i),$$

so from (2.1.8)

$$D(a) = a_0^{2n-2}(\det V)^2$$
$$= a_0^{2n-2} \det V^T V.$$

Equation (2.1.9) then follows by writing down $V^T V$.

2.1.4 From (2.1.3),

$$R_y(f\,g) = \begin{vmatrix} y+1 & 2y & y^3 & 0 \\ 0 & y+1 & 2y & y^3 \\ 0 & 1 & -6 & -3y^2 \\ 1 & -6 & -3y^2 & 0 \end{vmatrix}$$
$$= -y^4(4y + 3)^2.$$

$\therefore$ for $f = 0$, $g = 0$, we must have $y = 0$ or $y = -\frac{3}{4}$.

When $y = 0$, $x^2 - 6x = 0$ and $x^2 = 0$, so $x = 0$.

When $y = -\frac{3}{4}$,

$$\left.\begin{aligned} x^2/4 - 3x/2 - 27/64 &= 0 \\ x^2 - 6x - 27/16 &= 0 \end{aligned}\right\}$$

or $16x^2 - 96x - 27 = 0$ in both cases. The roots of this are $3 \pm \frac{3}{4}\sqrt{19}$.

2.1.5 Let $c(\lambda) = c_0\lambda^p + c_1\lambda^{p-1} + \cdots + c_p$. Then from (2.1.6),

$$\rho(a, c) = a_0^p c(\alpha_1) \cdots c(\alpha_n)$$
$$\rho(b, c) = b_0^p c(\alpha_1) \cdots c(\beta_m)$$

and

$$\rho(ab, c) = (a_0 b_0)^p c(\alpha_1) \cdots c(\alpha_n) c(\beta_1) \cdots c(\beta_m) = \rho(a, c)\rho(b, c)$$

since

$$a(\lambda)b(\lambda) = a_0 b_0 (\lambda - \alpha_1) \cdots (\lambda - \alpha_n)(\lambda - \beta_1) \cdots (\lambda - \beta_m).$$

2.1.6 From (2.1.8), $$D(a) = a_0^{2n-2} \prod_{i<j} (\alpha_i - \alpha_j)^2,$$
$$D(b) = b_0^{2m-2} \prod_{i<j} (\beta_i - \beta_j)^2,$$
$D(ab) =$
$$(a_0 b_0)^{2(n+m)-2} \prod_{i<j} (\alpha_i - \alpha_j)^2 \prod_{i<j} (\beta_i - \beta_j)^2 \prod_i \prod_j (\alpha_i - \beta_j)^2.$$
From (2.1.6),
$$(R(a, b))^2 = [a_0^m b_0^n \prod_i \prod_j (\alpha_i - \beta_j)]^2$$
$$= a_0^{2m} b_0^{2n} \prod_i \prod_j (\alpha_i - \beta_j)^2.$$
Result now follows.

2.1.7 From (2.1.8),
$$D(a) = a_0^{2n-2} \prod_{i<j} (\alpha_i - \alpha_j)^2.$$
For each pair of complex conjugate zeros $\theta \pm i\delta$ of $a(\lambda)$, $D(a)$ has a factor
$$(\theta + i\delta - \theta + i\delta)^2 = -4\delta^2 < 0.$$
For α_i, α_j both real, $(\alpha_i - \alpha_j)^2 > 0$. For α_i complex, α_j real, $D(a)$ has a pair of factors
$$(\alpha_i - \alpha_j)^2(\bar{\alpha}_i - \alpha_j)^2 = |\alpha_i - \alpha_j|^2 > 0.$$
For α_i, α_j both complex but $\alpha_i \neq \bar{\alpha}_j$, $D(a)$ has factors
$$(\alpha_i - \alpha_j)^2(\bar{\alpha}_i - \alpha_j)^2(\alpha_i - \bar{\alpha}_j)^2(\bar{\alpha}_i - \bar{\alpha}_j)^2 = |\alpha_i - \alpha_j|^2|\bar{\alpha}_i - \alpha_j|^2 > 0.$$
Thus $D(a)$ has sign $(-1)^k$, where k is the number of pairs of complex conjugate zeros of $a(\lambda)$.

2.2.1 $a'(\lambda) = a_1'(\lambda)a_2(\lambda) + a_1(\lambda)a_2'(\lambda)$, so $a'(A)$ is similar to
$$a_1'\begin{bmatrix} A_1 & 0 \\ 0 & A_2 \end{bmatrix} a_2 \begin{bmatrix} A_1 & 0 \\ 0 & A_2 \end{bmatrix} + a_1 \begin{bmatrix} A_1 & 0 \\ 0 & A_2 \end{bmatrix} a_2' \begin{bmatrix} A_1 & 0 \\ 0 & A_2 \end{bmatrix}$$
$$= \begin{bmatrix} a_1'(A_1)a_2(A_1) & 0 \\ 0 & a_1'(A_2)a_2(A_2) \end{bmatrix}$$
$$+ \begin{bmatrix} a_1(A_1)a_2'(A_1) & 0 \\ 0 & a_1(A_2)a_2'(A_2) \end{bmatrix}$$
$$= \begin{bmatrix} a_1'(A_1)a_2(A_1) & 0 \\ 0 & a_1(A_2)a_2'(A_2) \end{bmatrix}$$
since $a_1(A_1) = a_2(A_2) = 0$, by Cayley–Hamilton Theorem.

$\therefore$ det $a'(A) =$ det $a_1'(A_1)$ det $a_2(A_1)$ det $a_1(A_2)$ det $a_2'(A_2)$
$\Rightarrow D(a) = D(a_1)\{R(a_1, a_2)\}^2 D(a_2),$

since

$$\det a'(A) = D(a)/\varepsilon(n_1 + n_2),\ \det a_1'(A_1) = D(a_1)/\varepsilon(n_1),$$
$$\det a_2'(A_2) = D(a_2)/\varepsilon(n_2),\ \det a_2(A_1) = R(a_1, a_2)/\varepsilon(n_1),$$
$$\det a_1(A_2) = R(a_2, a_1)/\varepsilon(n_2) = (-1)^{n_1 n_2} R(a_1, a_2)/\varepsilon(n_1).$$

2.2.2 The characteristic roots of $a(A')$ are $a(z_i)$, where z_i are the characteristic roots of A', i.e. the roots of $a'(\lambda) = 0$, and so z_i are the stationary points of $a(\lambda)$. Thus the characteristic roots of $a(A')$ are the stationary values of $a(\lambda)$, and the largest is therefore the relative maximum.

2.2.3 From equation (2.2.3), $r_1 = [b_n, b_{n-1}, \ldots, b_2, b_1] = h$ and subsequent rows are obtained from (2.2.4) as $hA, hA^2, \ldots, hA^{n-1}$. Transposing gives required expression.

2.2.4 From (2.2.1) and (2.2.20),

$$f(\mu) = \det [(A + \mu I)^3 - 7(A + \mu I) + 6I]$$
$$= \begin{vmatrix} \mu^3 - 7\mu & 3\mu^2 & 3\mu \\ -18\mu & \mu^3 + 14\mu & 3\mu^2 \\ -18\mu^2 & 21\mu^2 - 18\mu & \mu^3 + 14\mu \end{vmatrix}$$
$$= \mu^3(\mu^6 - 42\mu^4 + 441\mu^2 - 400)$$
$$= \mu^3 g(\mu^2),$$

and $g(\nu)$ has desired properties.
Magnitude of discriminant $= 400$, being the coefficient of μ^3 in the expansion of the above determinant.

2.2.5 By Cayley–Hamilton, $A^n + a_1 A^{n-1} + \cdots + a_n I = 0$, so

$$R = b_0 A^n + b_1 A^{n-1} + \cdots + b_n I$$
$$= (b_1 - b_0 a_1)A^{n-1} + (b_2 - b_0 a_2)A^{n-2} + \cdots + (b_n - b_0 a_n)I$$
$$= r_{11}A^{n-1} + r_{12}A^{n-2} + \cdots + r_{1n}I.$$

2.2.6 $a(\lambda) = \lambda^4 + 2\lambda^3 + 2\lambda^2 + 2\lambda + 1,$

$$a'(A) = \begin{bmatrix} 2 & 4 & 6 & 4 \\ -4 & -6 & -4 & -2 \\ 2 & 0 & -2 & 0 \\ 0 & 2 & 0 & -2 \end{bmatrix} \quad \text{using (2.2.9) and (2.2.10).}$$

Rank $a'(A)$ is 3, and sequence (2.2.7) is 1, -4, -32, so $V = 1$.
$\therefore$ $a(\lambda)$ has 3 distinct zeros, and one distinct real zero [in fact $a(\lambda) = (\lambda + 1)^2(\lambda^2 + 1)$].

2.2.7 Let $d_0(\lambda) = (\lambda - \alpha_{11}) \cdots (\lambda - \alpha_{1p})(\lambda - \alpha_{21})^2 \cdots$
$$\times (\lambda - \alpha_{2q})^2(\lambda - \alpha_{31})^3 \cdots (\lambda - a_{41})^4 \cdots$$

then

$$d_1(\lambda) = (\lambda - \alpha_{21}) \cdots (\lambda - \alpha_{2q})(\lambda - \alpha_{31})^2 \cdots (\lambda - \alpha_{41})^3 \cdots$$
$$d_2(\lambda) = (\lambda - \alpha_{31}) \cdots (\lambda - \alpha_{41})^2 \cdots$$
$$d_3(\lambda) = (\lambda - \alpha_{41}) \cdots \text{etc.}$$

so

$$v_1(\lambda) = (\lambda - \alpha_{11}) \cdots (\lambda - \alpha_{1p})(\lambda - \alpha_{21}) \cdots$$
$$(\lambda - \alpha_{31}) \cdots (\lambda - \alpha_{41}) \cdots$$

$v_2(\lambda) = (\lambda - \alpha_{21}) \cdots (\lambda - \alpha_{31}) \cdots$
$v_3(\lambda) = (\lambda - \alpha_{31}) \cdots (\lambda - \alpha_{41}) \cdots$ etc.
$F_1(\lambda) = (\lambda - \alpha_{11}) \cdots (\lambda - \alpha_{1p})$, zeros of multiplicity one,
$F_2(\lambda) = (\lambda - \alpha_{21}) \cdots (\lambda - \alpha_{2q})$, zeros of multiplicity two, etc.

2.2.8 $a(\lambda) = (\lambda - \alpha_1)(\lambda - \alpha_2) \cdots (\lambda - \alpha_n)$, all α_i distinct.
The characteristic roots of $a'(A)$ are $a'(\alpha_i)$, so

$$\det a'(A) = a'(\alpha_1)a'(\alpha_2) \cdots a'(\alpha_n).$$

But

$$a'(\lambda) = a(\lambda)/(\lambda-\alpha_1)+a(\lambda)/(\lambda-\alpha_2)+ \cdots +a(\lambda)/(\lambda-\alpha_n),$$

so

$$\begin{aligned}\det a'(A) &= [(\alpha_1 - \alpha_2) \cdots (\alpha_1 - \alpha_n)][(\alpha_2 - \alpha_1) \cdots (\alpha_2 - \alpha_n)] \\ &\qquad \cdots [(\alpha_n - \alpha_1) \cdots (\alpha_n - \alpha_{n-1})] \\ &= (-1)^{1+2+\cdots+(n-1)} \prod_{i<j} (\alpha_i - \alpha_j)^2 \\ &= \varepsilon(n) \prod_{i<j} (\alpha_i - \alpha_j)^2 \\ &= \varepsilon(n) D(a) \text{ from (2.1.8).}\end{aligned}$$

2.3.1 $a_1 = 4, a_2 = 1, a_3 = -6$. From (2.3.2), $h_{11} = 8$,
$h_{12} = h_{21} = 0, h_{22} = 20, h_{13} = 12, h_{23} = 0, h_{33} = -12$.
$H_1 = 8, H_2 = 160, H_3 = -4800$, so by Theorem 2.12 $a(\lambda)$ has one zero in right half-plane and two in left.
From (2.3.6), $f(\theta) = \theta^3 - \theta + i(4\theta^2 + 6)$ and

$$F = \begin{bmatrix} 0 & 1 & 0 \\ 0 & 0 & 1 \\ 0 & 1 & 0 \end{bmatrix}$$

$$\begin{aligned} R(f) &= 4F^2 + 6I \\ &= \begin{bmatrix} 6 & 0 & 4 \\ 0 & 10 & 0 \\ 0 & 0 & 10 \end{bmatrix} \quad \text{(see proof of Theorem 2.13)} \end{aligned}$$

so sequence in Theorem 2.13 is 1, 4, 40, -600.
In fact $a(\lambda) = (\lambda - 1)(\lambda + 2)(\lambda + 3)$.

2.3.2 Suppose $a(\lambda)$ has a purely imaginary zero ik.
Then $f(\theta)$ has a real zero $-k$, so from (2.3.7) we have $\phi(-k) + i\psi(-k) = 0$, which implies $\phi(-k) = 0 = \psi(-k)$, i.e. that $\phi(\theta)$ and $\psi(\theta)$ have a common factor $\theta + k$. Thus $\phi(\theta)$, $\psi(\theta)$ relatively prime ensures that $a(\lambda)$ has no pure imaginary zero. The converse is not true. For example, if $a(\lambda) = (\lambda - 1)^2(\lambda + 1) = \lambda^3 - \lambda^2 - \lambda + 1$, then

$$\begin{aligned} f(\theta) &= \theta^3 - i\theta^2 + \theta - i \qquad \text{from (2.3.6)} \\ &= \theta(\theta^2 + 1) - i(\theta^2 + 1) \end{aligned}$$

so $\phi(\theta)$ and $\psi(\theta)$ are not relatively prime even though $a(\lambda)$ has no pure imaginary zero.

2.3.3 From (2.3.5), $f(\theta)$ has zeros iz, $-i\bar{z}$. In (2.3.7) $\phi(\theta)$ and $\psi(\theta)$ are both real, and we have

$$\phi(iz) + i\psi(iz) = 0$$
$$\phi(-i\bar{z}) + i\psi(-i\bar{z}) = 0.$$

Conjugate of first equation is $\phi(-i\bar{z}) - i\psi(-i\bar{z}) = 0$, and adding this to the second equation gives $\phi(-i\bar{z}) = \psi(-i\bar{z}) = 0$. That is, $\phi(\theta)$, $\psi(\theta)$ have common factor $\theta + i\bar{z}$, so R_n, which is resultant of $\phi(\theta)$ and $\psi(\theta)$, is zero.

2.3.4 $n = 3$:

$$\Delta_1 = a_1, \Delta_2 = a_1a_2 - a_3, \Delta_3 = a_3(a_1a_2 - a_3).$$

From (2.3.6), $f(\theta) = \theta^3 - a_2\theta + i(a_1\theta^2 - a_3)$, and

$$F = \begin{bmatrix} 0 & 1 & 0 \\ 0 & 0 & 1 \\ 0 & a_2 & 0 \end{bmatrix},$$

$$R = a_1F^2 - a_3I = \begin{bmatrix} -a_3 & 0 & a_1 \\ 0 & \begin{pmatrix} a_1a_2 \\ -a_3 \end{pmatrix} & 0 \\ 0 & 0 & \begin{pmatrix} a_1a_2 \\ -a_3 \end{pmatrix} \end{bmatrix}$$

and

$$R_1 = a_1 = \tfrac{1}{2}H_1 = \Delta_1;$$
$$\varepsilon(2)R_2 = a_1(a_1a_2 - a_3) = \tfrac{1}{4}H_2 = \Delta_1\Delta_2;$$
$$\varepsilon(3)R_3 = a_3(a_1a_2 - a_3)^2 = \tfrac{1}{8}H_3 = \Delta_2\Delta_3.$$

$n = 4$:

$$\Delta_1 = a_1, \Delta_2 = a_1a_2 - a_3,$$
$$\Delta_3 = a_3(a_1a_2 - a_3) - a_1^2a_4,$$
$$\Delta_4 = a_4[a_3(a_1a_2 - a_3) - a_1^2a_4].$$
$$f(\theta) = \theta^4 - a_2\theta^2 + a_4 + i(a_1\theta^3 - a_3\theta),$$

$$F = \begin{bmatrix} 0 & 1 & 0 & 0 \\ 0 & 0 & 1 & 0 \\ 0 & 0 & 0 & 1 \\ -a_4 & 0 & a_2 & 0 \end{bmatrix},$$

$$R = a_1F^3 - a_3F$$

$$= \begin{bmatrix} 0 & -a_3 & 0 & a_1 \\ -a_1a_4 & 0 & a_1a_2 - a_3 & 0 \\ 0 & -a_1a_4 & 0 & a_1a_2 - a_3 \\ -a_4(a_1a_2 - a_3) & 0 & \begin{pmatrix} -a_1a_4 \\ +a_2(a_1a_2 - a_3) \end{pmatrix} & 0 \end{bmatrix},$$

and

$$R_1 = a_1 = \tfrac{1}{2}H_1 = \Delta_1;$$
$$\varepsilon(2)R_2 = a_1(a_1a_2 - a_3) = \tfrac{1}{4}H_2 = \Delta_1\Delta_2;$$
$$\varepsilon(3)R_3 = a_3(a_1a_2 - a_3)^2 - a_1^2a_4(a_1a_2 - a_3) = \tfrac{1}{8}H_3 = \Delta_2\Delta_3;$$

$$\begin{aligned}\varepsilon(4)R_4 &= -a_4(a_1a_2 - a_3)[a_3(a_1a_2 - a_3)^2 - a_1^2a_4(a_1a_2 - a_3)] \\ &\quad + a_1a_4[a_2(a_1a_2 - a_3) - a_1a_4][a_3(a_1a_2 - a_3) - a_1^2a_4] \\ &= \tfrac{1}{16}H_4 = \Delta_3\Delta_4.\end{aligned}$$

2.3.5 Since $\tilde{g}(\eta) = \eta^n\bar{g}(1/\eta)$, the zeros of $\tilde{g}(\eta)$ are $1/\bar{z}_j$, where z_j are the zeros of $g(\eta)$. Thus if $g(\eta)$ has a zero $re^{i\theta}$, $\tilde{g}(\eta)$ has a zero $1/re^{-i\theta} = r^{-1}e^{i\theta}$, so $g(\eta)$ and $\tilde{g}(\eta)$ have a zero in common. Then $K_n = \det\tilde{g}(G)$, the resultant of $g(\eta)$ and $\tilde{g}(\eta)$, must be zero. This argument still applies for a zero on the unit circle ($r = 1$).

CHAPTER 3

3.1.1 The $\varepsilon_i(\lambda)$ are altered, but the $\psi_j(\lambda)$ are unchanged. This is because $g(\lambda)$ remains the same.

3.2.1 Let

$x(\lambda) = (\lambda - \lambda_1)^{\alpha_1} \cdots (\lambda - \lambda_r)^{\alpha_r}$, $y(\lambda) = (\lambda - \mu_1)^{\beta_1} \cdots (\lambda - \mu_s)^{\beta_s}$ with $\lambda_i \neq \mu_j$. Then $\delta[x/y] = \alpha_1 + \cdots + \alpha_r$ if $\sum\alpha_i > \sum\beta_j$; $= \beta_1 + \cdots + \beta_s$ if $\sum\alpha_i \leqslant \sum\beta_j$.

In Theorem 3.2, (i) clearly holds.

(ii) $\delta[x/y] = 0 \Rightarrow \sum\alpha_i = \sum\beta_j = 0 \Rightarrow x/y$ is independent of λ.
(iii) $\delta[g^{-1}] = \delta[y/x] = \delta[x/y]$ from the definition.
(iv) $\delta[g_1 + g_2] = \delta[x_1/y_1 + x_2/y_2] = \delta[(x_1y_2 + x_2y_1)/y_1y_2]$.

If g_1 and g_2 have no poles in common (including that at infinity), then

$$\begin{aligned}\delta[g_1 + g_2] &= \max[\delta(x_1y_2 + x_2y_1), \delta(y_1y_2)] \\ &= \max[\delta(x_1) + \delta(y_2), \delta(x_2) + \delta(y_1), \delta(y_1) + \delta(y_2)]\end{aligned}$$

and

$$\delta[g_1] + \delta[g_2] = \max[\delta(x_1), \delta(y_1)] + \max[\delta(x_2), \delta(y_2)],$$

and these are equal in all possible cases.

If g_1 and g_2 have pole(s) in common, factors will cancel in $(g_1 + g_2)$, so $\delta[g_1 + g_2] < \delta[g_1] + \delta[g_2]$.

(v) $\delta[g_1g_2] = \delta[x_1x_2/y_1y_2] = \max[\delta(x_1) + \delta(x_2), \delta(y_1) + \delta(y_2)]$ and $\delta[g_1] + \delta[g_2]$ is as in (iv). Again equality except when cancellation(s) occur.

3.2.2 Obvious from $G(\lambda) = C\,\mathrm{Adj}\,(\lambda I - A)B/\det(\lambda I - A)$ (provided no cancellations).

3.2.3 Using A in equation (2.2.1), $A^{(1)} = A + a_1I$ has last column $[0, 0, \ldots, 0, 1, 0]^T$, $A^{(2)} = AA^{(1)} + a_2I$ has last column $[0, 0, \ldots, 0, 1, 0, 0]^T, \ldots, A^{(k)}$ has as last column the $(n - k)$th column of I_n.

3.2.4 $$C(\lambda I - A)^{-1}B = [\beta_n, \ldots, \beta_1] \operatorname{Adj}(\lambda I - A)\begin{bmatrix}0\\0\\ \cdot\\ \cdot\\ \cdot\\ 0\\ 1\end{bmatrix} \div \det(\lambda I - A)$$

$$= [\beta_n, \ldots, \beta_1][\text{last column of } \operatorname{Adj}(\lambda I - A)] \div (\lambda^n + a_1\lambda^{n-1} + \cdots + a_n)$$
$$= (\beta_n + \beta_{n-1}\lambda + \cdots + \beta_1\lambda^{n-1})/(\lambda^n + a_1\lambda^{n-1} + \cdots + a_n),$$

using Exercise 3.2.3. Minimal since $\delta[G(\lambda)] = n$ (Exercise 3.2.1).

3.2.5 As in previous exercise

$$C(\lambda I - A)^{-1}B = C[\text{last column of } \operatorname{Adj}(\lambda I - A)] \div \det(\lambda I - A)$$
$$= \begin{bmatrix}\beta_{1n} + \cdots + \beta_{11}\lambda^{n-1}\\ \vdots \\ \beta_{mn} + \cdots + \beta_{m1}\lambda^{n-1}\end{bmatrix} \div (\lambda^n + a_1\lambda^{n-1} + \cdots + a_n).$$

Easy to verify controllable and observable (see proof of Theorem 2.11) so realization is minimal.

3.2.6 $$(\lambda I - A)^{-1} = \begin{bmatrix}\lambda I - A_1 & -B_1C_2\\ 0 & \lambda I - A_2\end{bmatrix}^{-1}$$
$$= \begin{bmatrix}(\lambda I - A_1)^{-1} & (\lambda I - A_1)^{-1}B_1C_2(\lambda I - A_2)^{-1}\\ 0 & (\lambda I - A_2)^{-1}\end{bmatrix}$$

so

$$[C_1\ 0](\lambda I - A)^{-1}\begin{bmatrix}0\\ B_2\end{bmatrix}$$
$$= [C_1(\lambda I - A_1)^{-1}, C_1(\lambda I - A_1)^{-1}B_1C_2(\lambda I - A_2)^{-1}]\begin{bmatrix}0\\ B_2\end{bmatrix}$$
$$= C_1(\lambda I - A_1)^{-1}B_1C_2(\lambda I - A_2)^{-1}B_2$$
$$= G_1(\lambda)G_2(\lambda).$$

3.2.7 The (ij) element of $a_k\beta_k^T$ is

$$\lambda^{i-1}(\lambda^{x-j} + \lambda^{x-j-1}\tau_1 + \cdots + \tau_{x-j})$$
$$= \lambda^{x+i-j-1} + \lambda^{x+i-j-2}\tau_1 + \cdots + \lambda^{i-1}\tau_{x-j}.$$

If $i \geqslant j + 1$, this element is replaced in $(\alpha_k\beta_k^T)_M$ by the remainder after division by $\psi_k(\lambda)$, and this is

$$-(\tau_{x-j+1}\lambda^{i-2} + \tau_{x-j+2}\lambda^{i-3} + \cdots + \tau_n\lambda^{i-j-1}).$$

Straightforward multiplication then establishes that

$$(\alpha_k\beta_k^T)_M(\lambda I_x - A_k) = \psi_k(\lambda)I_x.$$

3.2.8 From equations (3.2.11) and (3.2.12),

$$[B_1, A_1B_1, A_1^2B_1] = \begin{bmatrix}1 & 1 & -2 & \\ -2 & -1 & 11 & \ldots\\ 11 & 7 & -22 & \end{bmatrix}$$

which has rank 3, so controllable.

$$[C_1^T, A_1^T C_1^T, (A_1^T)^2 C_1^T] = \begin{bmatrix} 1 & 0 & 0 & 0 & \\ 0 & -1 & 1 & 0 & \dots \\ 0 & 0 & 0 & -1 & \end{bmatrix},$$

rank 3, so observable.

3.2.9

$$K_1 = \lim_{\lambda \to 1} (\lambda - 1)G(\lambda) = \tfrac{1}{6}\begin{bmatrix} 7 & 6 \\ -7 & -6 \end{bmatrix}, r_1 = 1$$

$$K_2 = \lim_{\lambda \to -1} (\lambda + 1)G(\lambda) = -\tfrac{1}{2}\begin{bmatrix} 7 & 4 \\ 7 & 4 \end{bmatrix}, r_2 = 1$$

$$K_3 = \lim_{\lambda \to -2} (\lambda + 2)G(\lambda) = \tfrac{1}{3}\begin{bmatrix} 10 & 6 \\ 20 & 12 \end{bmatrix}, r_3 = 1$$

$$K_1 = L_1 M_1 = \begin{bmatrix} 1 \\ -1 \end{bmatrix} [\tfrac{7}{6} \quad 1]$$

$$K_2 = L_2 M_2 = \begin{bmatrix} 1 \\ 1 \end{bmatrix} [-\tfrac{7}{2} \quad -2]$$

$$K_3 = L_3 M_3 = \begin{bmatrix} 1 \\ 2 \end{bmatrix} [\tfrac{10}{3} \quad 2].$$

Minimal realization is $A_2 = \text{diag}\,[1, -1, -2]$,

$$B_2 = \begin{bmatrix} \tfrac{7}{6} & 1 \\ -\tfrac{7}{2} & -2 \\ \tfrac{10}{3} & 2 \end{bmatrix}, \qquad C_2 = \begin{bmatrix} 1 & 1 & 1 \\ -1 & 1 & 2 \end{bmatrix}.$$

T can be determined from

$C_2 = C_1 T^{-1}$, $T^{-1}B_2 = B_1$, $T^{-1}A_2 = A_1 T^{-1}$ (Theorem 3.4).

The first two rows of T^{-1} are obtained at once from the first equation, the first two elements of the last row of T^{-1} follow from the third equation, and the (33) element of T^{-1} is then obtained by using the second equation. Inversion gives

$$6T = \begin{bmatrix} 2 & 3 & 1 \\ 6 & -3 & -3 \\ -2 & 0 & 2 \end{bmatrix}.$$

3.2.10

$$gG = \begin{bmatrix} \lambda + 2 & 2(\lambda + 2) \\ -1 & \lambda + 1 \end{bmatrix}$$

$$\sim \begin{bmatrix} 1 & 0 \\ 0 & (\lambda + 2)(\lambda + 3) \end{bmatrix} \quad \text{(see Theorem 1.5)}$$

$\therefore$ by Theorem 3.1, McMillan form is

$$\begin{bmatrix} \dfrac{1}{(\lambda + 2)(\lambda + 1)} & 0 \\ 0 & \dfrac{\lambda + 3}{\lambda + 1} \end{bmatrix}.$$

$\therefore$ $\delta[G(\lambda)] = \delta(\psi_1) + \delta(\psi_2) = 2 + 1 = 3$, by Theorem 3.3.

Alternatively,

$$K_1 = \lim_{\lambda\to -1} (\lambda + 1)G(\lambda) = \begin{bmatrix} 1 & 2 \\ -1 & 0 \end{bmatrix}, r_1 = 2$$

$$K_2 = \lim_{\lambda\to -2} (\lambda + 2)G(\lambda) = \begin{bmatrix} 0 & 0 \\ 1 & 1 \end{bmatrix}, r_2 = 1$$

so by Theorem 3.8, $\delta[G] = 3$.

3.3.1 $C(\lambda I - A)^{-1}B$

$$= [C_{11}\, C_{12}] \begin{bmatrix} \lambda I_p - A_{11} & 0 \\ -A_{21} & \lambda I_{n-p} - A_{22} \end{bmatrix}^{-1} \begin{bmatrix} 0 \\ B_{21} \end{bmatrix}$$

$$= [C_{11}\, C_{12}] \begin{bmatrix} (\lambda I_p - A_{11})^{-1} & 0 \\ (\lambda I_{n-p} - A_{22})^{-1} A_{21} (\lambda I_p - A_{11})^{-1} & (\lambda I_{n-p} - A_{22})^{-1} \end{bmatrix} \times \begin{bmatrix} 0 \\ B_{21} \end{bmatrix}$$

$$= C_{12}(\lambda I_{n-p} - A_{22})^{-1}B_{21}.$$

3.3.2 First part follows as in solution to Exercise 3.2.3, or by considering the identity

$$(\text{Adj}\,(\lambda I - A_i))(\lambda I - A_i) = g_i(\lambda)I.$$

Since $C_i(\lambda I_{p(i)} - A_i)^{-1} = \begin{bmatrix} 0 \\ \cdot \\ \cdot \\ \cdot \\ 0 \\ \text{last row of } (\lambda I_{p(i)} - A_i)^{-1} \\ \cdot \\ \cdot \\ \cdot \\ 0 \end{bmatrix} \leftarrow \text{row } i,$

$$[C_1, C_2, \ldots, C_m] \begin{bmatrix} (\lambda I_{p(1)} - A_1)^{-1} & & & \\ & \cdot & & \\ & & \cdot & \\ & & & \cdot \\ & & & (\lambda I_{p(m)} - A_m)^{-1} \end{bmatrix} \times \begin{bmatrix} B_1 \\ \cdot \\ \cdot \\ \cdot \\ B_m \end{bmatrix}$$

$$= \begin{bmatrix} (1, \lambda, \ldots, \lambda^{p(1)-1})/g_1 & 0 & \cdot\ \cdot\ \cdot \\ 0 & (1, \lambda, \ldots, \lambda^{p(2)-1})/g_2 & \cdot\ \cdot\ \cdot \\ \cdot & \cdot & \\ \cdot & & \cdot \\ \cdot & & & \cdot \end{bmatrix} \begin{bmatrix} B_1 \\ \cdot \\ \cdot \\ \cdot \\ B_m \end{bmatrix}$$

$$= \begin{bmatrix} u_1/g_1 \\ \cdot \\ \cdot \\ \cdot \\ u_m/g_m \end{bmatrix}.$$

3.3.3 Using Theorem 3.12, least common denominator of all minors of $G(\lambda)$ is $g(\lambda)$, so $\nu(G) = \delta g = 3$.

In Theorem 3.13, $g_2 = 2, g_1 = -1, g_0 = -2$,

$$H_2 = \begin{bmatrix} 1 & 1 \\ 2 & 1 \end{bmatrix}, \quad H_1 = \begin{bmatrix} 0 & 1 \\ -7 & -5 \end{bmatrix}, \quad H_0 = \begin{bmatrix} 6 & 4 \\ -2 & -2 \end{bmatrix},$$

$Q_{10} = 0,\ Q_{11} = H_0,\ Q_{12} = H_1,\ Q_{13} = H_2$
$Q_{20} = 0,\ Q_{21} = 2H_2,\ Q_{22} = H_0 + H_2,\ Q_{23} = H_1 - 2H_2$
$Q_{30} = 0,\ Q_{31} = 2(H_1 - 2H_2),\ Q_{32} = H_1,$
$Q_{33} = H_0 - 2H_1 + 5H_2.$

The 6×6 matrix $[Q_{ij}], i, j = 1, 2, 3$ can be shown to have rank 3, by some standard method (for example using equivalence transformations).

3.3.4 Consider the product of $(\lambda I_{pl} - A)$ and the last block column of $(\lambda I_{pl} - A)^{-1}$:

$$\begin{bmatrix} \lambda I_l & -I_l & & & \\ & \lambda I_l & & & \\ & & \cdot & & \\ & & & \cdot & \\ & & & \cdot & -I_l \\ g_0 I_l & g_1 I_l & & \cdots & (\lambda + g_{p-1}) I_l \end{bmatrix} \begin{bmatrix} X_1 \\ \cdot \\ \cdot \\ \cdot \\ \cdot \\ X_p \end{bmatrix} = \begin{bmatrix} 0 \\ \cdot \\ \cdot \\ \cdot \\ 0 \\ I_l \end{bmatrix}.$$

This gives

$$X_2 = \lambda X_1, X_3 = \lambda X_2, \ldots, X_p = \lambda X_{p-1}$$

and

$$g_0 X_1 + g_1 X_2 + \cdots + (\lambda + g_{p-1}) X_p = I_p,$$

whence

$$X_1 = I_l/g(\lambda), X_i = \lambda^{i-1} X_1, i = 2, 3, \ldots, p.$$

$$\therefore\ C(\lambda I_{pl} - A)^{-1} B = [H_0, H_1, \ldots, H_{p-1}] \begin{bmatrix} X_1 \\ \cdot \\ \cdot \\ \cdot \\ X_p \end{bmatrix} \div g(\lambda)$$

$$= H(\lambda)/g(\lambda).$$

3.3.5 If $\begin{bmatrix} T & U \\ -V & W \end{bmatrix}$ corresponds to G then $\begin{bmatrix} T & U \\ -V & W + G_2 \end{bmatrix}$ corresponds to $G + G_2$.

The Smith form of T is the same for each system matrix, so by Theorem 3.15, the $\psi_i(\lambda)$ are unaltered.

3.3.6
$$P \sim \begin{bmatrix} I & 0 & 0 \\ 0 & I & T_1 \\ 0 & 0 & I \end{bmatrix} \begin{bmatrix} I & 0 & 0 \\ 0 & T_1 & U_1 \\ 0 & -I & D \end{bmatrix}$$
$$= \begin{bmatrix} I & 0 & 0 \\ 0 & 0 & U_1 + T_1 D \\ 0 & -I & D \end{bmatrix}$$
$$\sim \begin{bmatrix} I & & \\ & I & \\ & & U_1 + T_1 D \end{bmatrix},$$
so by Theorem 3.15, the ε_i are the invariant polynomials of $U_1 + T_1 D$.

Similarly
$$P \sim \begin{bmatrix} I & 0 & 0 \\ 0 & T_2 & I \\ 0 & -V_2 & D \end{bmatrix} \begin{bmatrix} I & 0 & 0 \\ 0 & I & 0 \\ 0 & -T_2 & I \end{bmatrix}$$
$$\sim \begin{bmatrix} I & & \\ & I & \\ & & V_2 + DT_2 \end{bmatrix}.$$

3.3.7
$$[V_1 \quad V_2] \begin{bmatrix} T_1 & 0 \\ 0 & T_2 \end{bmatrix}^{-1} \begin{bmatrix} U_1 \\ U_2 \end{bmatrix} + (W_1 + W_2)$$
$$= [V_1 \quad V_2] \begin{bmatrix} T_1^{-1} & 0 \\ 0 & T_2^{-1} \end{bmatrix} \begin{bmatrix} U_1 \\ U_2 \end{bmatrix} + (W_1 + W_2)$$
$$= V_1 T_1^{-1} U_1 + V_2 T_2^{-1} U_2 + W_1 + W_2 = G_1 + G_2.$$

3.3.8
$$D(\lambda^{-1}) = D_0/\lambda^s + \cdots + D_{s-1}/\lambda + D_s$$
$$= (D_0 + D_1\lambda + \cdots + D_{s-1}\lambda^{s-1})/\lambda^s + D_s.$$

In Theorem 3.13,

$Q_{10} = 0, \quad Q_{11} = D_0, \quad Q_{12} = D_1, \quad \ldots, \quad Q_{1s} = D_{s-1}$
$Q_{20} = 0, \quad Q_{21} = Q_{10}, \quad Q_{22} = Q_{11}, \quad Q_{23} = Q_{12}, \quad \ldots$
$Q_{30} = 0, \quad Q_{31} = Q_{20}, \quad Q_{32} = Q_{21}, \quad Q_{33} = Q_{22}, \quad \ldots$ etc.

and
$$[Q_{ij}] = \begin{bmatrix} D_0 & D_1 & \cdot & \cdot & \cdot & D_{s-1} \\ 0 & D_0 & & & & D_{s-2} \\ \cdot & & & \cdot & & \cdot \\ \cdot & & & & \cdot & \cdot \\ \cdot & & & & \cdot & \cdot \\ 0 & 0 & \cdot & \cdot & \cdot & D_0 \end{bmatrix},$$

$\nu(D(\lambda^{-1})) = \text{rank } [Q_{ij}]$.

3.4.1 Clearly conditions (i) and (ii) of the definition will be satisfied in each case.

For (iii):
$$x^*[Z_1^T(\bar{\lambda}) + Z_1(\lambda)]x \geq 0$$
$$x^*[Z_2^T(\bar{\lambda}) + Z_2(\lambda)]x \geq 0$$
$$\therefore \quad x^*[(Z_1^T(\bar{\lambda}))^T + Z_1^T(\lambda)]x = x^*[Z_1(\bar{\lambda}) + Z_1^T(\lambda)]x$$
$$= x^*[Z_1(\mu) + Z_1^T(\bar{\mu})]x$$
$$\geq 0 \text{ since } \operatorname{Re}(\mu) = \operatorname{Re}(\lambda) > 0.$$

Also

$$x^*[(Z_1(\bar{\lambda}) + Z_2(\bar{\lambda}))^T + Z_1(\lambda) + Z_2(\lambda)]x \geqslant 0,$$

and

$$x^*[R^T(Z_1^T(\bar{\lambda}) + Z_1(\lambda))R]x \geqslant 0.$$

3.4.2 $Y^T(-\lambda) = Z^T(-\lambda) + Z(\lambda) = Y(\lambda)$, so parahermitian. When $\lambda = i\omega$,

$$\begin{aligned} Y^*(\lambda) &= Y^T(\bar{\lambda}) \quad \text{since real-rational} \\ &= Y(-\bar{\lambda}) \quad \text{since parahermitian} \\ &= Z(-\bar{\lambda}) + Z^T(\bar{\lambda}) \\ &= Z(i\omega) + Z^T(-i\omega) = Y(i\omega) \end{aligned}$$

so $Y(i\omega)$ is Hermitian and positive semidefinite by (iv) of Theorem 3.18.

3.4.3

$$\begin{aligned} W_1^T(-\lambda)W_1(\lambda) &= W^T(-\lambda)V^T(-\lambda)V(\lambda)W(\lambda) \\ &= W^T(-\lambda)W(\lambda) = Y(\lambda). \end{aligned}$$

3.4.4 $Y^T(-\lambda) = A^T(-\lambda)A(\lambda) = Y(\lambda)$, so $Y(\lambda)$ is parahermitian, having rank r, and $Y(i\omega) = A^T(-i\omega)A(i\omega) = A^*(i\omega)A(i\omega)$ since A is real-rational, so $Y(i\omega)$ is positive semidefinite. Therefore by Theorem 3.23, $A(\lambda) = V(\lambda)W(\lambda)$.

3.4.5 Step (1) is obvious. For (2), det $Y_1(\lambda) = (-\lambda + 1)(\lambda + 1)$, so

$$T_2(\lambda) = \begin{bmatrix} 1 & 0 \\ 0 & \dfrac{1}{\lambda + 1} \end{bmatrix},$$

giving

$$T_2(-\lambda)Y_1(\lambda)T_2^T(\lambda) = \begin{bmatrix} -2\lambda^2 + 5 & \dfrac{-2\lambda^2 - 5\lambda + 13}{\lambda + 1} \\ \dfrac{-2\lambda^2 + 5\lambda + 13}{-\lambda + 1} & \dfrac{-2\lambda^2 + 34}{(-\lambda + 1)(\lambda + 1)} \end{bmatrix}.$$

Take

$$T_3(-\lambda) = \begin{bmatrix} 1 & 0 \\ \dfrac{k}{-\lambda + 1} & 1 \end{bmatrix},$$

and

$[(-2\lambda^2 + 5)k + (-2\lambda^2 + 5\lambda + 13)]_{\lambda=1} = 0$, giving $k = -16/3$.

(3)

$$Y_2^{(1)} = \begin{bmatrix} -2\lambda^2 & 26\lambda/3 \\ -26\lambda/3 & (26)^2/18 \end{bmatrix}$$

and

$$T_4(-\lambda) = \begin{bmatrix} 1 & \dfrac{-3\lambda}{13} \\ 0 & 1 \end{bmatrix}.$$

(4) $Y_3 = Y_4Y_4^T$ with Y_4 lower triangular.

3.4.6 $Z(\lambda) = C(\lambda I - A)^{-1}B = B^TP(\lambda I - A)^{-1}B$ from (3.4.5), and $Z^T(\bar{\lambda}) = B^T(\bar{\lambda}I - A^T)^{-1}PB$.

$$\begin{aligned}
\therefore\ & Z^T(\bar{\lambda}) + Z(\lambda) \\
&= B^T[(\bar{\lambda}I - A^T)^{-1}P + P(\lambda I - A)^{-1}]B \\
&= B^T(\bar{\lambda}I - A^T)^{-1}[P(\lambda I - A) + (\bar{\lambda}I - A^T)P](\lambda I - A)^{-1}B \\
&= B^T(\bar{\lambda}I - A^T)^{-1}[P(\lambda + \bar{\lambda}) + H^TH](\lambda I - A)^{-1}B \text{ using (3.4.5)} \\
&= [H(\lambda I - A)^{-1}B]^*[H(\lambda I - A)^{-1}B] \\
&\qquad + [(\lambda I - A)^{-1}B]^*P(\lambda + \bar{\lambda})[(\lambda I - A)^{-1}B],
\end{aligned}$$

which is positive semidefinite for $\text{Re}(\lambda) > 0$, since A is a stability matrix by virtue of Theorem 3.21.

3.4.7 $X = (Z + I)^{-1}(Z - I)$ gives $Z = (I + X)(I - X)^{-1}$.

$$\begin{aligned}
Z^*(i\omega) + Z(i\omega) &= Z^T(-i\omega) + Z(i\omega) \text{ since } Z \text{ real-rational} \\
&= (I - X^1)^{-1}(I + X^1) + (I + X)(I - X)^{-1}
\end{aligned}$$

where

$$X^1 = X^T(-i\omega) = X^*(i\omega).$$

$$\begin{aligned}
\therefore\ & Z^*(i\omega) + Z(i\omega) \\
&= (I - X^1)^{-1}[(I + X^1)(I - X) + (I - X^1)(I + X)](I - X)^{-1} \\
&= 2(I - X^1)^{-1}(I - X^1X)(I - X)^{-1}.
\end{aligned}$$

Also

$$I - X = (Z + I)^{-1}(Z + I - Z + I) = 2(Z + I)^{-1},$$

so

$$(I - X)^{-1} = \tfrac{1}{2}(Z + I) \text{ and } (I - X^1)^{-1} = \tfrac{1}{2}(Z^1 + I).$$

$$\begin{aligned}
\therefore\quad Z^*(i\omega) + Z(i\omega) &= \tfrac{1}{2}(I + Z^1)(I - X^1X)(I + Z) \\
&= \tfrac{1}{2}(I + Z(i\omega))^*(I - X^1X)(I + Z(i\omega)),
\end{aligned}$$

so if $Z^*(i\omega) + Z(i\omega)$ is positive semidefinite, so is $I - X^*(i\omega)X(i\omega)$ (provided $i\omega$ is not a pole of $Z(\lambda)$).

3.4.8

$$\begin{aligned}
&\{I + Y_0^{-1}(P_1B - C^T)^T[\lambda I - A - BY_0^{-1}(P_1B - C^T)^T]^{-1}B\} \\
&\quad \times Y_1^{-1}Y_1[I - Y_0^{-1}(P_1B - C^T)^T(\lambda I - A)^{-1}B] \\
&= I + Y_0^{-1}(P_1B - C^T)^T\{[\lambda I - A - BY_0^{-1}(P_1B - C^T)^T]^{-1} \\
&\quad \times(\lambda I - A - BY_0^{-1}(P_1B - C^T)^T) - I\}(\lambda I - A)^{-1}B \\
&= I.
\end{aligned}$$

3.4.9 From (3.4.9),

$$\begin{aligned}
A^TP_1 + P_1A &= (C^TY_0^{-1} - P_1BY_0^{-1})(B^TP_1 - C) \\
&= C^TY_0^{-1}B^TP_1 - C^TY_0^{-1}C - P_1BY_0^{-1}B^TP_1 + P_1BY_0^{-1}C.
\end{aligned}$$

$$\therefore\ P_1BY_0^{-1}B^TP_1 + (A^T - C^TY_0^{-1}B^T)P_1 + P_1(A - BY_0^{-1}C) + C^TY_0^{-1}C = 0,$$

or

$$P_1BY_0^{-1}B^TP_1 + \alpha^TP_1 + P_1\alpha + C^TY_0^{-1}C = 0.$$

3.4.10 Write $S(\lambda) = I - R(\lambda)$ and consider the definition at the beginning of Section 3.4.:

(i) is clearly satisfied.

(ii) $\bar{S}(\lambda) = I - \bar{R}(\lambda) = I - R(\bar{\lambda}) = S(\bar{\lambda})$.

(iii)
$$\begin{aligned}
I - R^T(\bar{\lambda})R(\lambda) &= I - (I - S^T(\bar{\lambda}))(I - S(\lambda)) \\
&= S^T(\bar{\lambda}) + S(\lambda) - S^T(\bar{\lambda})S(\lambda),
\end{aligned}$$

which is positive semidefinite by definition of $R(\lambda)$.

$\therefore$ since $S^T(\bar{\lambda})S(\lambda) = S^*(\lambda)S(\lambda)$ is positive semidefinite, so is $S^T(\bar{\lambda}) + S(\lambda)$.

3.4.11 The first two parts of the positive-real definition are obviously satisfied. For (iii), since

$$P^T(\bar{\lambda}) + P(\lambda) = \begin{bmatrix} (\lambda + \bar{\lambda})I - (A + A^T) & B - C^T \\ -C + B^T & D + D^T \end{bmatrix}$$

is positive semidefinite in Re $(\lambda) > 0$, so is $(\lambda + \bar{\lambda})I - (A + A^T)$, being a principal submatrix.

CHAPTER 4

4.1.1 Let $P = \text{diag}\,[p_1, p_2, \ldots, p_n]$. Equation (4.1.2) gives $p_1 - p_2 w_1 = 0, p_2 - p_3 w_2 = 0, \ldots, p_{n-1} - p_n w_{n-1} = 0,$ $2p_n w_n = 1$

so

$p_n = 1/2w_n, p_{n-1} = w_{n-1}/2w_n, \ldots, p_2 = w_2 w_3 \cdots w_{n-1}/2w_n,$ $p_1 = w_1 w_2 \cdots w_{n-1}/2w_n.$

By Theorem 4.2, W will be a stability matrix if and only if all $w_i > 0$.

4.1.2 In $T^T P + PT = -Q$, let $P = \text{diag}\,[p_1, \ldots, p_n]$ and $Q = \text{diag}\,[q_1, \ldots, q_n]$.

We then obtain

$$2p_1 t_{11} = -q_1, 2p_2 t_{22} = -q_2, \ldots, 2p_n t_{nn} = -q_n$$

and

$$p_1 t_{12} + p_2 t_{21} = 0, p_2 t_{23} + p_3 t_{32} = 0, \ldots, p_{n-1} t_{n-1,n} + p_n t_{n,n-1} = 0.$$

$\therefore$ take $q_1 = -2t_{11}, p_1 = 1,$

and

$$p_2 = -t_{12}/t_{21}, p_3 = t_{12}t_{23}/t_{21}t_{32}, \ldots, p_n = (-1)^{n-1} t_{12} t_{23} \cdots t_{n-1,n}/t_{21}t_{32} \cdots t_{n,n-1}$$

giving

$$q_2 = 2t_{12}t_{32}/t_{21}, \ldots, q_n = 2(-1)^n t_{12}t_{23} \cdots t_{n-1,n} t_{nn}/t_{21}t_{32} \cdots t_{n,n-1}.$$

If stated conditions hold, P and Q are positive definite so T is a stability matrix.

4.1.3 With $P = \begin{bmatrix} p_1 & p_2 \\ p_2 & p_3 \end{bmatrix}$, equation (4.1.2) gives

$$\begin{aligned} 8p_1 - 10p_2 \qquad\quad &= -\lambda_1 \\ 6p_1 - 3p_2 - 5p_3 &= 0 \\ 12p_2 - 14p_3 &= -\lambda_2, \end{aligned}$$

so that

$$p_1 = (51\lambda_1 + 25\lambda_2)/12, p_2 = (42\lambda_1 + 20\lambda_2)/12, p_3 = (36\lambda_1 + 18\lambda_2)/12.$$

With $\lambda_1 > 0$, $\lambda_2 > 0$, P is positive definite so A is a stability matrix by Theorem 4.2.
From (4.1.8),

$$B = \alpha \begin{bmatrix} 42\lambda_1 + 20\lambda_2 & 36\lambda_1 + 18\lambda_2 \\ -51\lambda_1 - 25\lambda_2 & -42\lambda_1 - 20\lambda_2 \end{bmatrix} - q_1 \begin{bmatrix} 51\lambda_1 + 25\lambda_2 & 42\lambda_1 + 20\lambda_2 \\ 0 & 0 \end{bmatrix} - q_2 \begin{bmatrix} 0 & 0 \\ 42\lambda_1 + 20\lambda_2 & 36\lambda_1 + 18\lambda_2 \end{bmatrix}.$$

4.1.4 $\operatorname{Tr} P_0S_0 = \operatorname{tr}(P_0S_0)^T = -\operatorname{tr}(S_0P_0) = -\operatorname{tr}(P_0S_0)$
$\therefore\ \operatorname{tr}(P_0S_0) = 0.$
From (4.1.7),

$$\Sigma\lambda_i(A) = \operatorname{tr}(A) = \operatorname{tr}(P_0S_0 - P_0Q_0) = -\operatorname{tr}(P_0Q).$$

Similarly in (4.1.8), $\operatorname{tr} B = -\operatorname{tr} Q_1P = -\Sigma\lambda_i(Q_1P)$. Since P is positive definite, Q_1 positive semidefinite, the characteristic roots of Q_1P are all nonnegative, and will have zero sum only when $Q_1 \equiv 0$. Thus if $Q_1 \not\equiv 0$, $\operatorname{tr} B \neq 0$, so principal diagonal of B cannot be identically zero.

4.1.5 $Ax_i = \lambda_i x_i$, so $x_i^*A^T = \bar{\lambda}_i x_i^*$ since A is real.
By (4.1.2),

$$x_i^*A^TPx_i + x_i^*PAx_i = -x_i^*Qx_i$$

so that

$$\bar{\lambda}_i x_i^*Px_i + \lambda_i x_i^*Px_i = -x_i^*Qx_i$$

$$\therefore \quad -x_i^*Qx_i/x_i^*Px_i = \bar{\lambda}_i + \lambda_i = 2\alpha_i.$$

Since P and Q are both positive definite symmetric, the characteristic roots of $\frac{1}{2}QP^{-1}$ are all real and positive, and by Rayleigh's theorem

$$\mu_1 \leqslant \tfrac{1}{2}x_i^*Qx_i/x_i^*Px_i \leqslant \mu_2.$$

$\therefore\ \mu_1 \leqslant -\alpha_i \leqslant \mu_2$, $i = 1, 2, \ldots, n$, so in particular $\mu_1 \leqslant -\alpha_1$, $-\alpha_n \leqslant \mu_2$.

4.1.6 Comparing (4.1.7) and (4.1.2) we have $P^{-1} = P_0$, $Q = 2Q_0$, so in the result of the previous exercise, the roots of $\frac{1}{2}QP^{-1}$ become the roots of Q_0P_0, which are the same as the roots of P_0Q_0. Thus the real parts of the characteristic roots of A in (4.1.7) lie between the greatest and least roots of the matrix $-P_0Q_0$ $(-\mu_1 \geqslant \alpha_i \geqslant \mu_2)$.

4.1.7 Let

$$\begin{bmatrix} T_1 & T_3 \\ T_4 & T_2 \end{bmatrix}\begin{bmatrix} A & 0 \\ 0 & -A^T \end{bmatrix} = \begin{bmatrix} A & 0 \\ 0 & -A^T \end{bmatrix}\begin{bmatrix} T_1 & T_3 \\ T_4 & T_2 \end{bmatrix}$$

so $-T_3A^T = AT_3$, $T_4A = -A^TT_4$. Since A is a stability matrix, A and $-A^T$ have no characteristic roots in common, so Theorem 1.9 implies that $T_3 = 0 = T_4$. Also, $T_1A = AT_1$ and $-T_2A^T = -A^TT_2$.

4.1.8 Replace A in Theorem 4.1 by $A - kI$, which has characteristic roots $\lambda_i(A) - k$. Equation (4.1.2) becomes $A^TP + PA - 2kP = -Q$, which has positive definite solution P if and only if $\lambda_i(A) - k < 0$.

4.1.9 $P - \alpha^TP\alpha = M + \alpha^TM\alpha + (\alpha^T)^2M\alpha^2 + \cdots$
$$-\alpha^TM\alpha - (\alpha^T)^2M\alpha^2 - \cdots = M.$$
If $\quad L^{-1}\alpha L = \Lambda = \operatorname{diag}[\lambda_1, \lambda_2, \ldots, \lambda_n]$, then
$$(\alpha^T)^rM\alpha^r = (L^T)^{-1}\Lambda^rL^TMLA^rL^{-1}$$
$$= (L^T)^{-1}\Lambda^rK\Lambda^rL^{-1}, \text{ where } K = L^TML = [k_{ij}],$$
and $[\Lambda^rK\Lambda^r]_{ij} = k_{ij}\lambda_j^r\lambda_j^r$.
$$P = (L^T)^{-1}[L^TML + \Lambda K\Lambda + \Lambda^2K\Lambda^2 + \cdots]L^{-1},$$
and
$$[L^TML + \Lambda K\Lambda + \Lambda^2K\Lambda^2 + \cdots]_{ij}$$
$$= k_{ij}(1 + \lambda_i\lambda_j + \lambda_i^2\lambda_j^2 + \cdots)$$
$$= k_{ij}(1 - \lambda_i\lambda_j)^{-1},$$
which converges since $|\lambda_i|\,|\lambda_j| < 1$ if α is a convergent matrix.

4.1.10 Differentiating $\dot{x} = Ax$ with respect to θ gives $\dot{y} = A_\theta x + Ay$, so $\dot{X} = BX$ where
$$X = \begin{bmatrix} x \\ y \end{bmatrix}, \qquad B = \begin{bmatrix} A & 0 \\ A_\theta & A \end{bmatrix}.$$
Det $(\lambda I_{2n} - B) = \{\det(\lambda I_n - A)\}^2$, so if A is a stability matrix when $\theta = \theta_0$, so is B, and by Theorem 4.1 $X \to 0$ as $t \to \infty$, $\Rightarrow y \to 0$ as $t \to \infty$. Similarly, if $z = (\partial^2x/\partial\theta^2)$,
$\dot{Z} = A_{\theta\theta}x + 2A_\theta y + Az$, $\dot{Y} = CY$ where
$$Y = \begin{bmatrix} x \\ y \\ z \end{bmatrix}, \qquad C = \begin{bmatrix} A & 0 & 0 \\ A_\theta & A & 0 \\ A_{\theta\theta} & A_\theta & A \end{bmatrix}$$
and the argument proceeds as before.

4.1.11 $y = Tx$, $\dot{y} = TAT^{-1}x + Tbu$.
$$\text{Let } T^{-1} = [s_1, \ldots, s_n], \text{ so } TAT^{-1} = \begin{bmatrix} t \\ tA \\ \cdot \\ \cdot \\ \cdot \\ tA^{n-1} \end{bmatrix} [As_1, \ldots, As_n].$$
$$TT^{-1} = I_n = \begin{bmatrix} t \\ tA \\ \cdot \\ \cdot \\ \cdot \\ tA^{n-1} \end{bmatrix} [s_1, \ldots, s_n]$$
$$\Rightarrow TAT^{-1} = \begin{bmatrix} 0 & 1 & & & & \\ 0 & 0 & 1 & & & \\ \cdot & \cdot & \cdot & \cdot & \cdot & \cdot \\ & & & & & 1 \\ f_1 & f_2 & \cdot & \cdot & \cdot & f_n \end{bmatrix} = F, \quad \text{where } f_i = tA^ns_i.$$

Also

$$Tb = \begin{bmatrix} t \\ tA \\ \vdots \\ tA^{n-1} \end{bmatrix} b = \begin{bmatrix} 0 \\ 0 \\ \vdots \\ 0 \\ 1 \end{bmatrix} = d, \text{ say.}$$

$\therefore\ tb = 0,\ tAb = 0, \ldots, tA^{n-1}\,b = 1$, or

$$t[b, Ab, \ldots, A^{n-1}b] = d^T.$$

If $R[b, Ab, \ldots, A^{n-1}b] = n$, clearly a nonzero vector t exists. Conversely, if T exists,

Rank $[b, Ab, \ldots, A^{n-1}b]$

$$= R[Tb, TAb, \ldots, TA^{n-1}b]$$

$$= R[Tb, (TAT^{-1})Tb, (TAT^{-1})^2Tb, \ldots, (TAT^{-1})^{n-1}Tb]$$

$$= \begin{bmatrix} 0 & 0 & \cdot & \cdot & \cdot & \cdot & 0 & 1 \\ \cdot & & & & & \cdot & & \\ \cdot & & & & \cdot & & & \\ \cdot & & & \cdot & & & & \\ & & 1 & & & & & \\ 0 & 1 & & & & & & \\ 1 & f_n & \cdot & \cdot & \cdot & \cdot & \cdot & \cdot \end{bmatrix}$$

$$= n.$$

4.1.12 $\dot{y} = Fy + du = (F + dk)y$ where $u = ky$ and

$$k = [k_1, \ldots, k_n].$$

Also

$$dk = \begin{bmatrix} & 0 & \\ k_1 & \cdots & k_n \end{bmatrix},$$

so $F + dk$ is in companion form with last row

$$[k_1 + f_1, k_2 + f_2, \ldots, k_n + f_n].$$

Since the k_i can be chosen arbitrarily, $F + dk$ can have arbitrary characteristic equation and thus arbitrary characteristic roots. Finally,

$$F + dk = TAT^{-1} + Tbk = T(A + bkT)T^{-1},$$

so $K = kT$ satisfies the conditions of Theorem 4.8.

4.1.13 $d(x^TPx)/dt = \dot{x}^TPx + x^TP\dot{x} = x^T(A^TP + PA)x = x^TQx$, using (4.1.1). Integrating with respect to t,

$$\int_0^t x^TQx\,dt = x^T(t)Px(t) - x^T(0)Px(0).$$

If A is a stability matrix, $x(t) \to 0$ as $t \to \infty$.

4.1.14 From (4.1.7), $A^TP_0^{-1} + P_0^{-1}A = -2Q_0$.

$$\therefore \text{ by previous exercise, } \int_0^\infty x^TQ_0x\,dt = \tfrac{1}{2}x^T(0)P_0^{-1}x(0).$$

4.1.15
$$\begin{aligned} d(x \otimes x)/dt &= \dot{x} \otimes x + x \otimes \dot{x} \\ &= Ax \otimes x + x \otimes Ax \text{ using (4.1.1)} \\ &= (A \otimes I + I \otimes A)(x \otimes x). \end{aligned}$$

Integrating with respect to t from 0 to ∞, using $x(\infty) = 0$, gives

$$-x(0) \otimes x(0) = B \int_0^\infty x \otimes x \, dt.$$

4.1.16 $d(tx^TPx)/dt = x^TPx + tx^TQx$, as in Exercise 4.1.13.

$\therefore$ integration gives $\displaystyle\int_0^\infty tx^TQx \, dt = \Big[tx^TPx\Big]_0^\infty - \int_0^\infty x^TPx \, dt$

$$= x^T(0)P_1x(0),$$

where $A^TP_1 + P_1A = P$, as in Exercise 4.1.13.
Similarly, $d(t^2x^TP_1x)/dt = 2tx^TP_1x + t^2x^TQx$, and

$$\begin{aligned} \int_0^\infty t^2x^TQx \, dt &= -2 \int_0^\infty tx^TP_1x \, dt \\ &= -2x^T(0)P_2x(0), \text{ where } A^TP_2 + P_2A = P_1, \end{aligned}$$

etc.

4.1.17 $d(tx \otimes x)/dt = x \otimes x + tBx \otimes x$, as in Exercise 4.1.15.

$$\begin{aligned} \therefore \int_0^\infty tx \otimes x \, dt &= B^{-1}\Big[tx \otimes x\Big]_0^\infty - B^{-1}\int_0^\infty x \otimes x \, dt \\ &= B^{-2}x(0) \otimes x(0). \end{aligned}$$

Similarly, $d(t^2x \otimes x)/dt = 2tx \otimes x + t^2Bx \otimes x$,

$$\begin{aligned} \int_0^\infty t^2x \otimes x \, dt &= B^{-1}\Big[t^2x \otimes x\Big]_0^\infty - 2B^{-1}\int_0^\infty tx \otimes x \, dt \\ &= -2B^{-3}x(0) \otimes x(0), \end{aligned}$$

etc.

4.1.18 $Ax_i = \lambda_i x_i$, $x_i^*A^T = x_i^*\bar{\lambda}_i$ since A is real.

$\therefore\ x_i^*A^TQx_i - x_i^*QAx_i = \bar{\lambda}_i x_i^*Qx_i - \lambda_i x_i^*Qx_i = 0,$

so $\bar{\lambda}_i = \lambda_i$, all i, since $x_i^*Qx_i > 0$.

4.1.19
$$\begin{vmatrix} \lambda + 4 & -3e^{-8t} \\ e^{8t} & \lambda \end{vmatrix} = \lambda^2 + 4\lambda + 3 = (\lambda + 1)(\lambda + 3).$$

$$\dot{x}_1 = -4x_1 + 3e^{-8t}x_2,\ \dot{x}_2 = -e^{8t}x_1$$

give

$$e^{8t}(\dot{x}_1 + 4x_1) = 3x_2,$$

or, differentiating, $8e^{8t}(\dot{x}_1 + 4x_1) + e^{8t}(\ddot{x}_1 + 4\dot{x}_1) = 3(-e^{8t}x_1)$.

$$\therefore\ \ddot{x}_1 + 12\dot{x}_1 + 35x_1 = 0,$$

which has general solution $x_1 = Ae^{-7t} + Be^{-5t}$. This leads to

$$x_2 = -Ae^t - \tfrac{1}{3}Be^{3t} \to \infty \text{ as } t \to \infty.$$

4.1.20 $\alpha^TR^{-1}\alpha - R^{-1}$

$$= (T^T - R)^{-1}(T^T + R)R^{-1}(T + R)(T - R)^{-1} - R^{-1}$$

$$= (T^T - R)^{-1}[(T^TR^{-1} + I)(T + R) - (T^TR^{-1} - I)(T - R)](T - R)^{-1}$$
$$= (T^T - R)^{-1}[2(T + T^T)](T - R)^{-1},$$

which is negative definite if $T + T^T$ is negative definite, in which case the given matrix is convergent, by Theorem 4.7, provided R (and thus R^{-1}) is positive definite.

4.2.1 If $H = \text{diag}\,[h_1, \ldots, h_n]$, for C in (4.2.1) to be diagonal, $a_{i,i+1}h_i + \bar{a}_{i+1,i}\,h_{i+1} = 0$, $i = 1, 2, \ldots, n - 1$, since the h_i are real. $\therefore a_{i,i+1} = -h_{i+1}\bar{a}_{i+1,i}/h_i = k_i\bar{a}_{i+1,i}$.

4.2.2 $(HA)^*H^{-1} + H^{-1}(HA) = A^* + A$, which is positive definite.
$\therefore$ by Theorem 4.11, $\text{In}\,(HA) = \text{In}\,(H^{-1}) = \text{In}\,(H)$.

4.2.3 Let $\text{In}\,(H_1) = (m, n - m, 0) = \text{In}\,(H_1^{-1})$ since H_1 is Hermitian and nonsingular. By Theorem 4.14,

$$\begin{aligned}\text{In}\,(H) &= \text{In}\,(H_1) + \text{In}\,(-H_2^*H_1^{-1}H_2)\\ &= \text{In}\,(H_1) + \text{In}\,(-H_1^{-1}) \text{ by Theorem 4.9}\\ &= (m, n - m, 0) + (n - m, m, 0) = (n, n, 0).\end{aligned}$$

4.2.4 $Ax = \lambda x$, $x^*A = \lambda x^*$ since $A^* = A$, λ real.
$\therefore x^*AHx + x^*HAx = 2\lambda x^*Hx = 0$
$\therefore \lambda = 0$ since $x^*Hx > 0$.
$\therefore A = 0$ since all its characteristic roots are zero.

4.2.5 From (4.2.1),

$$\begin{aligned}\alpha^2 - \beta^2 &= \tfrac{1}{2}(\alpha - \beta)(\alpha + \beta) + \tfrac{1}{2}(\alpha + \beta)(\alpha - \beta)\\ &= A^*H + HA, \quad \text{where } A = \alpha - \beta.\end{aligned}$$

$\therefore \text{In}\,(\alpha - \beta) = \text{In}\,H = \text{In}\,(\alpha + \beta) = (n, 0, 0)$ since α, β are both positive definite.

4.2.6 If $Hx = \lambda x$, then $x^*H = \lambda x^*$ since H is Hermitian.
$\therefore x^*(A^*H + HA)x = \lambda(x^*A^*x + x^*Ax) > 0$.
$\therefore \lambda \neq 0$, so H is nonsingular since it has no zero roots.

4.3.1 If $X = S - kI$, D positive diagonal, then

$$\begin{aligned}(XD)^TD + D(XD) &= (SD - kD)^TD + D(SD - KD)\\ &= -DSD - kD^2 + DSD - kD^2\\ &= -2kD^2, \text{ which is positive definite.}\end{aligned}$$

$\therefore$ by Theorem 4.2, XD is stable; i.e. X is D-stable.

4.3.2 Let A have $2k$ complex and $n - 2k$ real characteristic roots. For a complex pair $\alpha \pm i\beta$, product is $\alpha^2 + \beta^2 > 0$.
$\therefore$ Since $\det A$ = product of characteristic roots of A,

$$\text{sgn}\,(\det A) = (-1)^{n-2k} = (-1)^n,$$

since all the real roots of A are negative.
Also,

$$\det A = \sum_{j=1}^{n} a_{ij}A_{ij},$$

so each nonzero term in the sum must have sign of $(-1)^n$.
$\therefore$ sgn (a_{ij}) sgn $(A_{ij}) = (-1)^n$.

4.3.3 If $b = 0$, then det $B = 0 \Rightarrow$ there exists $y \neq 0$ such that $By = 0$. $\therefore B(-y) = 0$, contradiction, for if $y \in Q(x)$, then $-y \notin Q(x)$, so det $B \neq 0$.
If $b \neq 0$, $Bx_1 = b$, then if det $B = 0$ we have $B(x_1 + \lambda y) = b$ for arbitrary scalar λ, y being as above. By a suitable choice of λ, we can make $x_1 + \lambda y \notin Q(x)$, again a contradiction.

4.3.4 $[\lambda I - A]_{ii} = \lambda - a_{ii} = \alpha + i\beta - a_{ii}$, say.

$$\begin{aligned}\therefore\ |\lambda - a_{ii}|^2 &= \alpha^2 + \beta^2 + a_{ii}^2 - 2\alpha a_{ii} \\ &\geqslant a_{ii}^2 \text{ since } \alpha \geqslant 0,\ a_{ii} < 0.\end{aligned}$$

$$\therefore\ |\lambda - a_{ii}| \geqslant |a_{ii}| > \sum_{j \neq i} k_j |a_{ij}|/k_i \text{ by (4.3.1),}$$

so $\lambda I - A$ is also q.d.d.
By given result, $|\lambda I - A| \neq 0$ for Re $(\lambda) \geqslant 0$, so A can have no characteristic roots with nonnegative real parts.

4.3.5 *Gershgorin's Theorem*: Every characteristic root of A lies in at least one of the n discs with centres a_{ii}, radii

$$\rho_i = \sum_{j \neq i}^{n} |a_{ij}|.$$

From (4.3.1), A diagonal dominant implies $|a_{ii}| > \rho_i$, so if $a_{ii} < 0$ all the discs lie wholly within the left half-plane, and therefore so do all the characteristic roots of A.
If $B \in Q(A)$, and $a_{ii} < 0$, then any B having $|b_{ii}|$ large enough so that B is diagonal dominant will be a stability matrix.

CHAPTER 5

5.1.1 $\dot{w} = -\ddot{y}/ey + \dot{y}\dot{e}/e^2y + \dot{y}^2/ey^2$, and substitution into (5.1.2) gives

$$\ddot{y}e - \dot{y}(\dot{e} + fe) + ge^2y = 0.$$

5.1.2

$$\begin{aligned}\dot{w} &= \dot{y}_2/y_1 - y_2\dot{y}_1/y_1^2 \\ &= (gy_1 + \tfrac{1}{2}fy_2)/y_1 - y_2(-\tfrac{1}{2}fy_1 - ey_2)/y_1^2 \\ &= g + fy_2/y_1 + ey_2^2/y_1^2 \\ &= g + fw + ew^2.\end{aligned}$$

5.1.3

$$\begin{aligned}\dot{X} &= \dot{X}_1X_0X_2 + X_1X_0\dot{X}_2 \\ &= HX_1X_0X_2 + X_1X_0X_2K \\ &= HX + XK\end{aligned}$$

and $$X(t_0) = X_1(t_0)X_0X_2(t_0) = X_0.$$

5.1.4 In Theorem 5.1, $K(t) = -H(t)$, so from (5.1.4), $\dot{X}_2 = -X_2H$, $\dot{X}_1 = HX_1$ and hence $X_2 = X_1^{-1}$, since $d(X_1^{-1})/dt = -X_1^{-1}\dot{X}_1X_1^{-1} = -X_1^{-1}H$. Thus general solution is $X_1(t)X_0X_1^{-1}(t)$ where X_1 is given by (5.1.4).

5.1.5 $\dot{Y} = \ddot{X}X^{-1} - \dot{X}X^{-1}\dot{X}X^{-1} = K - Y^2$.

5.1.6 Let $U_1 = V_1^{-1}$, $U_2 = V_2^{-1}$ so that

$$\dot{U}_1 = -V_1^{-1}\dot{V}_1V_1^{-1} = -V_1^{-1}(V_1EV_1 + DV_1 + V_1F)V_1^{-1} = -E - U_1D - FU_1$$

and similarly

$$\dot{U}_2 = -E - U_2D - FU_2.$$

$$\therefore\ \dot{U}_2 - \dot{U}_1 = -(U_2 - U_1)D - F(U_2 - U_1),$$

so by Theorem 5.1,

$$\text{rank}\ [U_2(t) - U_1(t)] = \text{rank}\ [V_2^{-1}(t_0) - V_1^{-1}(t_0)] = \text{constant}.$$

5.1.7 $\dot{W} = \dot{Y}Z + Y\dot{Z} = AYZ - YZA = AW - WA$.

$$\therefore\ \text{Trace}\ (\dot{W}) = \text{trace}\ (AW - WA) = \text{trace}\ AW - \text{trace}\ WA = 0,$$

so trace $W(t)$ = constant.

5.1.8 $\dot{X} = A^TX + XA$ (i)

by Theorem 5.1 has solution $X(t) = X_1X_0X_2$ where

$$X_1 = \exp(A^Tt),\ X_2 = \exp(At).$$

Integrate (i) with respect to t from t_0 to ∞ to get

$$[X(\infty) - X(t_0)] = A^T\int_{t_0}^{\infty} X\,dt + \left(\int_{t_0}^{\infty} X\,dt\right)A.$$

Since A is a stability matrix, $X(\infty) = 0$ so that

$$-X_0 = A^TY + YA,$$

where

$$Y = \int_{t_0}^{\infty} X_1X_0X_2\,dt.$$

5.1.9 From Theorem 5.6, the solution of (5.1.1) is $W = ZY^{-1}$, so $Y^TWY = Y^TZ$ showing $W(t)$ is symmetric if $Y^T(t)Z(t)$ is symmetric. Let $K = Y^TZ - Z^TY$, then

$$\begin{aligned}\dot{K} &= \dot{Y}^TZ + Y^T\dot{Z} - \dot{Z}^TY - Z^T\dot{Y}\\ &= (-Y^TF^T - Z^TE^T)Z + Y^T(GY + DZ)\\ &\quad - (Y^TG^T + Z^TD^T)Y - Z^T(-FY - EZ) \quad \text{using (5.1.12)}\\ &= 0 \text{ since } F = D^T,\ E = E^T,\ G = G^T.\end{aligned}$$

$\therefore K(t)$ is a constant matrix and is in fact zero since when $t = t_s$, $W^T(t_s) = W(t_s)$, i.e. $Y^T(t_s)Z(t_s) = Z^T(t_s)Y(t_s)$. Therefore $Y^T(t)Z(t)$ is symmetric, all t.

5.2.1 $\dot{P} = [\dot{\phi}_3 + \dot{\phi}_4M][\phi_1 + \phi_2M]^{-1} - [\phi_3 + \phi_4M][\phi_1 + \phi_2M]^{-1}[\dot{\phi}_1 + \dot{\phi}_2M][\phi_1 + \phi_2M]^{-1}$

and

$$\dot{\phi}_1 = A\phi_1 - BR^{-1}B^T\phi_3,\ \dot{\phi}_2 = A\phi_2 - BR^{-1}B^T\phi_4$$

$$\dot{\phi}_3 = -Q\phi_1 - A^T\phi_3,\ \dot{\phi}_4 = -Q\phi_2 - A^T\phi_4, \text{ using (5.2.8).}$$

Substituting,

$$\dot{P} = (-Q\phi_1 - A^T\phi_3 - Q\phi_2 M - A^T\phi_4 M)(\phi_1 + \phi_2 M)^{-1} - (\phi_3 + \phi_4 M)(\phi_1 + \phi_2 M)^{-1} \times [A\phi_1 - BR^{-1}B^T\phi_3 + A\phi_2 M - BR^{-1}B^T\phi_4 M] \times [\phi_1 + \phi_2 M]^{-1}$$

$$= -Q - A^T(\phi_3 + \phi_4 M)(\phi_1 + \phi_2 M)^{-1} - (\phi_3 + \phi_4 M)(\phi_1 + \phi_2 M)^{-1} \times [A - BR^{-1}B^T(\phi_3 + \phi_4 M)(\phi_1 + \phi_2 M)^{-1}]$$

$$= -Q - A^TP - PA + PBR^{-1}B^TP.$$

$$P(t_1) = [\phi_3(t_1) + \phi_4(t_1)M][\phi_1(t_1) + \phi_2(t_1)M]^{-1}$$

$= M$, since $\phi_1(t_1) = \phi_4(t_1) = I_n$, $\phi_2(t_1) = \phi_3(t_1) = 0$.

5.2.2 $A^TP + PA \rightarrow (A^T - PS)P + P(A + SP) = A^TP + PA$, so (5.2.4) is unaltered.

5.2.3

$$\dot{P}_1 = \dot{P} - \dot{X}Y^{-1}X^T + XY^{-1}\dot{Y}Y^{-1}X^T - XY^{-1}\dot{X}^T$$
$$= (PBR^{-1}B^TP - A^TP - PA - Q) + (A^T - PBR^{-1}B^T)XY^{-1}X^T + XY^{-1}(X^TBR^{-1}B^TX)Y^{-1}X^T - XY^{-1}X^T(-A + BR^{-1}B^TP)$$
$$= (P - XY^{-1}X^T)BR^{-1}B^T(P - XY^{-1}X^T) - A^T(P - XY^{-1}X^T) - (P - XY^{-1}X^T)A - Q$$
$$= P_1BR^{-1}B^TP_1 - A^TP_1 - P_1A - Q.$$

5.2.4

$$x^TQ_1x + 2x^TSu_1 + u_1^TRu_1 = x^T(Q_1 - SR^{-1}S^T)x + x^T(Su_1 + SR^{-1}S^Tx) + u_1^T(Ru_1 + S^Tx)$$
$$= x^T(Q_1 - SR^{-1}S^T)x + x^TSR^{-1}(R)(u_1 + R^{-1}S^Tx) + u_1^TR(u_1 + R^{-1}S^Tx)$$
$$= x^TQx + u^TRu.$$

5.2.5

$$\mathcal{J}^T = \begin{bmatrix} 0 & I \\ -I & 0 \end{bmatrix} = -\mathcal{J}.$$

$$(-\mathcal{J})(\mathcal{J}) = \begin{bmatrix} 0 & I \\ -I & 0 \end{bmatrix}\begin{bmatrix} 0 & -I \\ I & 0 \end{bmatrix} = \begin{bmatrix} I & 0 \\ 0 & I \end{bmatrix} \text{ so } -\mathcal{J} = \mathcal{J}^{-1}.$$

$$\mathcal{J}^2 = \begin{bmatrix} 0 & -I \\ I & 0 \end{bmatrix}\begin{bmatrix} 0 & -I \\ I & 0 \end{bmatrix} = \begin{bmatrix} -I & 0 \\ 0 & -I \end{bmatrix} = -I_{2n}.$$

From (5.2.6),

$$\mathcal{J}H^T\mathcal{J} = \begin{bmatrix} 0 & -I \\ I & 0 \end{bmatrix}\begin{bmatrix} A^T & -Q \\ -BR^{-1}B^T & -A \end{bmatrix}\begin{bmatrix} 0 & -I \\ I & 0 \end{bmatrix}$$
$$= \begin{bmatrix} 0 & -I \\ I & 0 \end{bmatrix}\begin{bmatrix} -Q & -A^T \\ -A & BR^{-1}B^T \end{bmatrix}$$
$$= H.$$

5.2.6 $H\alpha_i = \lambda_i\alpha_i \quad \therefore \mathcal{J}H^T\mathcal{J}\alpha_i = \lambda_i\alpha_i.$

$\therefore H^T\mathcal{J}\alpha_i = -\lambda_i\mathcal{J}\alpha_i$ since $\mathcal{J}^{-1} = -\mathcal{J}$.

Transposing, $(\mathcal{J}\alpha_i)^TH = -\lambda_i(\mathcal{J}\alpha_i)^T$, so $-\lambda_i$ is a characteristic root with left characteristic vector $(\mathcal{J}\alpha_i)^T$.
Similarly using $H\beta_i = -\lambda_i\beta_i$.

5.2.7 By the assumption that all the characteristic roots of H are distinct, $L = [\alpha_1, \alpha_2, \ldots, \alpha_n, \beta_1, \beta_2, \ldots, \beta_n]$ is nonsingular and $L^{-1}HL = \text{diag}\,[\lambda_1, \ldots, \lambda_n, -\lambda_1, \ldots, -\lambda_n]$. By the previous exercise

$$L^{-1} = \begin{bmatrix} \beta_1^T \mathcal{J} \\ \cdot \\ \cdot \\ \cdot \\ \beta_n^T \mathcal{J} \\ -\alpha_1^T \mathcal{J} \\ \cdot \\ \cdot \\ \cdot \\ -\alpha_n^T \mathcal{J} \end{bmatrix} \quad \text{since } \mathcal{J}^T = -\mathcal{J}$$

$$= \begin{bmatrix} L_2^T \mathcal{J} \\ -L_1^T \mathcal{J} \end{bmatrix}.$$

5.2.8

$$\mathcal{J}^T L^T \mathcal{J} = \begin{bmatrix} 0 & I \\ -I & 0 \end{bmatrix} \begin{bmatrix} L_1^T \mathcal{J} \\ L_2^T \mathcal{J} \end{bmatrix}$$

$$= \begin{bmatrix} L_2^T \mathcal{J} \\ -L_1^T \mathcal{J} \end{bmatrix} = L^{-1}.$$

$$\therefore\ L^T \mathcal{J} L = \mathcal{J} \quad \text{since } (\mathcal{J}^T)^{-1} = \mathcal{J}.$$

5.2.9

$$\dot{\psi}^T = -\phi^{-1}\dot{\phi}\phi^{-1}$$
$$= -\phi^{-1}H = -\psi^T H.$$
$$\therefore\ \dot{\psi} = -H^T\psi,$$

so

$$\mathcal{J}^T \dot{\psi} \mathcal{J} = -\mathcal{J} H^T \psi \mathcal{J}$$
$$= H(\mathcal{J}^T \psi \mathcal{J}) \qquad \text{(ii)}$$

since $\mathcal{J} H^T \mathcal{J} = H \Rightarrow \mathcal{J} H^T = -H\mathcal{J} = H\mathcal{J}^T$. Hence, by virtue of (ii) and $\mathcal{J}^T \psi(t_0) \mathcal{J} = \mathcal{J}^T [\phi^{-1}(t_0)]^T \mathcal{J} = I_{2n}$, we have

$$\mathcal{J}^T \psi(t) \mathcal{J} = \phi(t). \qquad \text{(iii)}$$

Equation (iii) gives

$$\phi = \mathcal{J}^T (\phi^{-1})^T \mathcal{J}.$$
$$\therefore\ \mathcal{J}\phi = (\phi^{-1})^T \mathcal{J}, \quad \text{since } \mathcal{J}^T = \mathcal{J}^{-1}.$$
$$\therefore\ \phi^T \mathcal{J} \phi = \mathcal{J}.$$

5.3.1 Let P_1, P_2 be two solutions of (5.3.4), so that

$$P_1 B R^{-1} B^T P_1 - A^T P_1 - P_1 A - Q = 0$$
$$P_2 B R^{-1} B^T P_2 - A^T P_2 - P_2 A - Q = 0.$$

Subtract:

$$(P_1 - P_2)(A - BR^{-1}B^T P_1) + (A^T - P_2 B R^{-1} B^T)(P_1 - P_2) = 0. \qquad \text{(iv)}$$

By Theorem 5.13,

$$A_1 = A - BR^{-1}B^TP_1,\ A_2 = A - BR^{-1}B^TP_2$$

are stability matrices, so A_1 and $-A_2^T$ have no characteristic roots in common. Therefore, by Theorem 1.9, the solution to (iv) is unique, and must be zero, so $P_1 = P_2$.

5.3.2

$$H_0 = \begin{bmatrix} A & 0 \\ -Q & -A^T \end{bmatrix},$$

$$\det(\lambda I_{2n} - H_0) = \det(\lambda I_n - A)\det(\lambda I_n + A^T),$$

so characteristic roots of H_0 are $\mu_1, \ldots, \mu_n, -\mu_1, \ldots, -\mu_n$. In Theorem 5.15, solution of $A^TP + PA = -Q$ is

$$P = [c_1, \ldots, c_n][b_1, \ldots, b_n]^{-1}$$

where

$$\begin{bmatrix} A & 0 \\ -Q & -A^T \end{bmatrix}\begin{bmatrix} b_i \\ c_i \end{bmatrix} = \mu_i\begin{bmatrix} b_i \\ c_i \end{bmatrix}$$

so

$$Ab_i = \mu_i b_i$$

and

$$-Qb_i - A^Tc_i = \mu_i c_i.$$

$\therefore$ b_i is a characteristic vector of A, and

$$(A^T + \mu_i I)c_i = -Qb_i.$$

5.3.3 From the previous solution, $P = -(B^T)^{-1}P_1B^{-1}$ where

$$B = [b_1, \ldots, b_n] \text{ and } P_1 = -B^T[c_1, \ldots, c_n] = [p_{ij}].$$

$$\therefore\ p_{ij} = -b_i^Tc_j = b_i^T(A^T + \mu_j I)^{-1}Qb_j.$$

Since $Ab_i = \mu_i b_i$, we have

$$(A + \mu_j I)b_i = (\mu_i + \mu_j)b_i$$

so

$$b_i/(\mu_i + \mu_j) = (A + \mu_j I)^{-1}b_i$$

or

$$b_i^T/(\mu_i + \mu_j) = b_i^T(A^T + \mu_j I)^{-1}$$

$$\therefore\ p_{ij} = b_i^TQb_j/(\mu_i + \mu_j).$$

5.3.4 Since B is the matrix of characteristic vectors of A,

$$B^{-1}AB = \text{diag}\,[\mu_1, \ldots, \mu_n] = \Lambda.$$

On substitution $A^TP + PA = -Q$ becomes

$$(B^T)^{-1}\Lambda B^TP + PB\Lambda B^{-1} = -Q$$

or

$$\Lambda(B^TPB) + (B^TPB)\Lambda = -B^TQB.$$

Writing $B^TPB = X = [x_{ij}]$, this gives

$$\mu_i x_{ij} + x_{ij}\mu_j = -b_i^TQb_j$$

$$\therefore\ x_{ij} = -b_i^TQb_j/(\mu_i + \mu_j),$$

so $X = -P_1$, and $P = (B^T)^{-1}XB^{-1}$ as before.

5.3.5 The closed-loop system matrix is

$$A - BR^{-1}B^T[c_1, \ldots, c_n][b_1, \ldots, b_n]^{-1}$$

and

$$\{A - BR^{-1}B^T[c_1, \ldots, c_n][b_1, \ldots, b_n]^{-1}\}[b_1, \ldots, b_n]e_i$$
$$= [b_1, \ldots, b_n] \operatorname{diag} [\lambda_1, \ldots, \lambda_n]e_i \text{ using (5.3.21)}$$
$$= [b_1, \ldots, b_n]\lambda_i e_i,$$

so $[b_1, \ldots, b_n]e_i (= b_i)$ is a characteristic vector corresponding to λ_i.

5.3.6 From Exercise 1.1.1, the remainder on division of $A(\lambda)$ by $\lambda I - X$ on the left is

$$R = A_p + XA_{p-1} + \cdots + X^p A_0,$$

so if $\lambda I - A$ is a left factor $R = 0$; and conversely.

5.3.7 In Theorem 5.14, $E = BY_0^{-1}B^T$, $D = (A - BY_0^{-1}C)^T$, $F = A - BY_0^{-1}C$, $G = C^T Y_0^{-1}C$, and $P_1 = T_3 T_1^{-1}$. Equation (5.3.12) gives $(A - BY_0^{-1}C)T_1 + BY_0^{-1}B^T T_3 = -T_1\beta_1$.

$$\therefore\ (A - BY_0^{-1}C) + BY_0^{-1}B^T P_1 = -T_1\beta_1 T_1^{-1}$$

or

$$A + BY_0^{-1}(B^T P_1 - C) = T_1(-\beta_1)T_1^{-1}.$$

CHAPTER 6

6.1.1 If $A^*Ax = 0$ then $x^*A^*Ax = 0$, or $(Ax)^*Ax = 0$, so $Ax = 0$. Conversely, if $Ax = 0$ then $A^*Ax = 0$. The systems of equations $Ax = 0$, $A^*Ax = 0$ are therefore equivalent, so rank $A =$ rank A^*A.

6.1.2 $A[A^*(AA^*)^{-1}] = I_m$.

6.1.3 $A^+ = \operatorname{diag} [a_1^+, a_2^+, \ldots, a_n^+]$.
If H Hermitian, $H = U^*DU$ with U unitary, D diagonal so $H^+ = U^*D^+U$ by (vi), Theorem 6.2.

6.1.4 (i) Obvious.
(ii) Let $B = A^+$, then $BAB = B$, $ABA = A$, $(AB)^* = AB$, $(BA)^* = BA$ so $B^+ = A$.
(iii) Let $B = A^*$, easily verified that $B^+ = (A^+)^*$ using (6.1.6)–(6.1.9).
(iv) If $k \neq 0$, $(kA)^+ = (1/k)A^+$; if $k = 0$, $(kA)^+ = 0 = k^+A^+$.

6.1.5 By equivalence transformations, $J = PEQ$ where P, Q nonsingular and E diagonal with first k ($=$ rank J) elements unity, others zero. $J^2 = J \Rightarrow EQPE = E$, so trace $E = k =$ trace $(EQ)(PE) =$ trace $(PE)(EQ) =$ trace $PEQ =$ trace J.
$(A^+A)^2 = A^+AA^+A = A^+A$ so rank $A^+A =$ trace A^+A.
From (6.1.5), rank $A^+ = r =$ rank A, $A^+A = D^*(DD^*)^{-1}D$ so rank $A^+A = r =$ rank $A =$ rank A^*A (Exercise 6.1.1).

6.1.6 $A^*AA^+ = A^*(AA^+)^* = A^*(A^+)^*A^* = (AA^+A)^* = A^*$.
$A^+AA^* = (A^+A)^*A^* = (AA^+A)^* = A^*$.
$A^*(A^+)^*A^+ = (A^+A)^*A^+ = A^+AA^+ = A^+$.
$A^+(A^+)^*A^* = A^+(AA^+)^* = A^+AA^+ = A^+$.

6.1.7 $(I_n - A^+A)^* = I_n - (A^+A)^* = I_n - A^+A.$
$(I_n - A^+A)^2 = I_n - A^+A - A^+A + A^+AA^+A = I_n - A^+A.$
Similarly for $I_m - AA^+$.

6.1.8 $HHH = HH = H$, $(HH)^* = H^* = H$, so from (6.1.6)–(6.1.9), $H^+ = H$.

6.1.9 Let $B = (A^*A)^+A^*$. Then

$$ABA = A(A^*A)^+A^*A = AA^+(A^+)^*A^*A = AA^+(AA^+)^*A$$
$$= AA^+AA^+A = AA^+A = A.$$
$$BAB = (A^*A)^+A^*A(A^*A)^+A^* = (A^*A)^+A^* = B.$$
$$(AB)^* = [A(A^*A)^+A^*]^* = A(A^*A)^+A^* = AB.$$
$$(BA)^* = [(A^*A)^+A^*A]^* = A^*A(A^*A)^+ = A^*AA^+(A^+)^*$$
$$= A^*(A^+)^* \quad \text{by Exercise 6.1.6}$$
$$= (A^+A)^* = A^+A$$
$$= A^+(A^+)^*A^*A \quad \text{by Exercise 6.1.6}$$
$$= (A^*A)^+A^*A = BA.$$

6.1.10 $A^*AA^+(A^+)^*A^+ = DA^*AA^*AA^+(A^+)^*A^+$
Left side $= A^*(A^+)^*A^+ = A^+$ using Exercise 6.1.6.
Right side $= DA^*AA^*(A^+)^*A^+ = DA^*AA^+ = DA^*$ using Exercise 6.1.6.

6.1.11 From (6.1.5), $A^+ = (A^*A)^{-1}A^*$, $B^+ = B^*(BB^*)^{-1}$,
$(AB)^+ = B^*(BB^*)^{-1}(A^*A)^{-1}A^* = B^+A^+.$

6.1.12 If $k = 0$, $(AA^*)^m = 0 \Rightarrow A = 0$ since trace $AA^* = \sum\sum |a_{ij}|^2$.

6.2.1 $A(B_1AB_2)A = AB_2A = A.$
$(B_1AB_2)A(B_1AB_2) = B_1AB_2AB_2 = B_1AB_2.$

6.2.2 (i) $B^2 = A^{(1)}AA^{(1)}A = A^{(1)}A = B.$
(ii) As in Exercise 6.1.5, trace B = rank B. Since $A = AB$, rank $A \leqslant$ rank B, and $B = A^{(1)}AB$ so rank $B \leqslant$ rank A. $\therefore$ rank A = rank B.
(iii) $A(A^{(1)}Y + BZ - Z) = AA^{(1)}Y + AA^{(1)}AZ - AZ = AA^{(1)}Y = Y$ since $AX = Y \Rightarrow AA^{(1)}Y = AA^{(1)}AX = AX = Y$ if consistent.

6.2.3 $A = AA^{(1)}A$ so rank $A \leqslant$ rank $A^{(1)}$. If $A^{(1)} \in A(1, 2)$,
$A^{(1)} = A^{(1)}AA^{(1)}$ so rank $A^{(1)} \leqslant$ rank A $\therefore$ rank A = rank $A^{(1)}$.

6.2.4 If $ABA = A$, then
$(C^*C)^{-1}C^*ABAD^*(DD^*)^{-1} = (C^*C)^{-1}C^*AD^*(DD^*)^{-1}$
$\Rightarrow DBC = I_r$ using $A = CD$.
Conversely, if $DBC = I_r$ then $CDBCD = CD$ or $ABA = A$.

6.2.5 Using $AX = XA$, $X(A^p + m_1A^{p+1} + \cdots + m_kA^{p+k}) = 0$ reduces to

$$AXA(A^{p-2} + m_1A^{p-1} + \cdots + m_kA^{p+k-2}) = 0$$

or
$$A^{p-1} + m_1A^p + \cdots + m_kA^{p+k-1} = 0,$$

provided $p \geqslant 2$.

Contradiction unless $p = 0$ or 1. If $p = 0$,

$$I_n + m_1 A + \cdots + m_k A^k = 0 \Rightarrow X = A^{-1} = -m_1 I \cdots -m_k A^{k-1}.$$

If $p = 1$,

$$A + m_1 A^2 + \cdots + m_k A^{k+1} = 0 \Rightarrow XA = -m_1 A \cdots -m_k A^k$$

and

$$\begin{aligned} X = XAX &= -m_1 AX - \cdots -m_k A^k X \\ &= -m_1 AX - m_2 A \cdots -m_k A^{k-1} \\ &= \text{polynomial in } A. \end{aligned}$$

6.2.6 $(BA - CA)(BA - CA)^* = BAA^*B^* - BAA^*C^* - CAA^*B^* + CAA^*C^* = 0$,

$$\|BA - CA\|^2 = \text{trace}\,(BA - CA)(BA - CA)^* = 0 \Rightarrow BA - CA = 0.$$

$$\therefore\ AA^*(AA^*)^{(1)}AA^* = AA^* \Rightarrow AA^*(AA^*)^{(1)}A = A.$$

(i) $A(A^{(1)}AA^{(1)})A = AA^{(1)}A = A$;
$(A^{(1)}AA^{(1)})A(A^{(1)}AA^{(1)}) = A^{(1)}AA^{(1)}AA^{(1)} = A^{(1)}AA^{(1)}$.

(ii) $A[A^*(AA^*)^{(1)}]A = A$;
$[A^*(AA^*)^{(1)}]A[A^*(AA^*)^{(1)}] = A^*(AA^*)^{(1)}$;
$[A^*(AA^*)^{(1)}A]^* = A^*(AA^*)^{(1)}A^*$.

6.2.7 $A^*AA^*(A^*AA^*)^{(1)}A^*AA^* = A^*AA^*$

$\Rightarrow AA^*(A^*AA^*)^{(1)}A^*A = A$ by Exercise 6.2.6
$\Rightarrow A^+AA^*(A^*AA^*)^{(1)}A^*AA^+ = A^+AA^+$
$\Rightarrow A^*(A^*AA^*)^{(1)}A^* = A^+$, by Exercise 6.1.6.

6.2.8

$$\begin{aligned} AXA &= P^{-1}\begin{bmatrix} B & 0 \\ 0 & 0 \end{bmatrix} Q^{-1}Q \begin{bmatrix} Z & U \\ V & W \end{bmatrix} PP^{-1} \begin{bmatrix} B & 0 \\ 0 & 0 \end{bmatrix} Q^{-1} \\ &= P^{-1}\begin{bmatrix} BZB & 0 \\ 0 & 0 \end{bmatrix} Q^{-1} \\ &= A \text{ if and only if } BZB = B \Rightarrow Z = B^{-1}. \end{aligned}$$

$$\begin{aligned} XAX &= Q\begin{bmatrix} ZBZ & ZBU \\ VBZ & VBU \end{bmatrix} P \\ &= Q\begin{bmatrix} Z & U \\ V & VBU \end{bmatrix} P \quad \text{since } Z = B^{-1} \\ &= X \text{ if and only if } VBU = W. \end{aligned}$$

6.2.9

$$AX = P^*\begin{bmatrix} BZ & BU \\ 0 & 0 \end{bmatrix} P \quad \text{since } P^{-1} = P^*.$$

$$AXA = A \Leftrightarrow Z = B^{-1},\ XAX = X \Leftrightarrow W = VBU,$$

$(AX)^* = AX \Rightarrow BU = 0 \Rightarrow U = 0 \Rightarrow W = 0$ and conversely.
Similarly, if in addition $(XA)^* = XA$ then $V = 0$.

6.2.10 $BA^{r+1} = (BA^r)A = A^rBA = A^{r-1}(ABA) = A^r$ ($=A^{r+1}B$, similarly).

$$Ax = \lambda x \Rightarrow BA^{r+1}x = \lambda BA^r x \Rightarrow A^r x = \lambda^{r+1}Bx \Rightarrow \lambda^{-1}x = Bx.$$

Conversely,

$$By = \mu y \Rightarrow A^{r+1}By = \mu A^{r+1}y$$
$$\Rightarrow A^r y = \mu A^{r+1}y \Rightarrow \mu^{-1}(A^r y) = A(A^r y).$$

6.2.11 $$AA_1 = AA_2AA_1 = A_2AA_1A = A_2A = AA_2.$$

If $A_1, A_2 \in A(1, 2)$ then

$$A_1 = A_1AA_1 = A_2AA_1 = A_2AA_2 = A_2.$$

6.2.12 $$Ax = \lambda x \Rightarrow AA^RAx = \lambda AA^Rx \Rightarrow Ax = \lambda A^RAx$$
$$\Rightarrow \lambda x = \lambda^2 A^Rx \Rightarrow \lambda^{-1}x = A^Rx.$$

Conversely,

$$A^Ry = \mu y \Rightarrow A^RAA^Ry = \mu A^RAy \Rightarrow A^Ry = \mu AA^Ry$$
$$\Rightarrow \mu y = \mu^2 Ay \Rightarrow \mu^{-1}y = Ay.$$

6.2.13 $E^2 = ABAB = AB = E$. $\therefore$ $\dot{E} = \dot{E}E + E\dot{E}$, so

$$E\dot{E} = E\dot{E}E + E^2\dot{E} = E\dot{E}E + E\dot{E} \Rightarrow E\dot{E}E = 0.$$

6.2.14 $B_2 = B_2D_1 = D_2B_2$, so if $\dot{D}_2B_2 = B_2\dot{D}_1 = 0$ then

$$\dot{B}_2 = \dot{B}_2D_1 = D_2\dot{B}_2 \text{ so } -B_2\dot{A}B_2 = D_2\dot{B}_2D_1 = \dot{B}_2D_1 = \dot{B}_2.$$

Conversely, if $\dot{B}_2 = -B_2\dot{A}B_2$ then $\dot{B}_2 = D_2\dot{B}_2D_1$

$$\text{so } D_2\dot{B}_2 = D_2^2\dot{B}_2D_1 = D_2\dot{B}_2D_1 = \dot{B}_2.$$

Similarly $\dot{B}_2 = \dot{B}_2D_1$ so $\dot{D}_2B_2 = B_2\dot{D}_1 = 0$.

6.3.1 $$B(I + \gamma CB)(A + \beta C) = BA + \beta BC + \gamma BCBA + \gamma\beta BCBC$$
$$= I + \beta BC + \gamma BC + \gamma\beta\alpha BC$$
$$= I \text{ provided } \beta + \gamma + \alpha\beta\gamma = 0.$$

CHAPTER 7

7.1.1 Suitable matrices (not of course unique) are

$$P = \begin{bmatrix} 1 & 0 & 0 \\ 2 & -1 & 0 \\ -1 & 2 & -1 \end{bmatrix}, \quad Q = \begin{bmatrix} 1 & 0 & 1 & -1 \\ 0 & 1 & 2 & -2 \\ 0 & 0 & 1 & 3 \\ 0 & 0 & 0 & 1 \end{bmatrix}$$

and $$S = \begin{bmatrix} 1 & & & \\ & 1 & & \\ & & 1 & 0 \end{bmatrix}.$$

$$Ax = b \Rightarrow Sy = Pb, x = Qy,$$

giving

$$y_1 = b_1, y_2 = 2b_1 - b_2, y_3 = -b_1 + 2b_2 - b_3, y_4 = t$$

so

$$x_1 = 2b_2 - b_3 - t, x_2 = 3b_2 - 2b_3 - 2t,$$
$$x_3 = -b_1 + 2b_2 - b_3 + 3t, x_4 = t,$$

where t is an arbitrary integer.

7.1.2 If A is prime, see Theorem 7.2.

If $PAQ = [S, 0]$ then mth determinantal divisor of A is $d = s_1 s_2 \ldots s_m$. $Ax = b \Rightarrow Sy = c$, $x = Qy$, so y integral $\Leftrightarrow x$ integral. If an integral solution y exists, then $s_1 y_1 = c_1$, $s_2 y_2 = c_2, \ldots$, which implies $[S, c] \sim [S, 0]$. But

$$[A, b] \sim P[A, b] \begin{bmatrix} Q & 0 \\ 0 & 1 \end{bmatrix} = [S, Pb] = [S, c] \sim [S, 0]$$

so mth determinantal divisor of $[A, b]$ = divisor of $[S, 0] = d$. Conversely, if mth determinantal divisor of $[A, b]$ is d, this implies mth divisor of $[S, c]$ is also d.

i.e.

$$\text{g.c.d.}\,(c_1 s_2 \ldots s_m,\ c_2 s_1 s_3 \ldots s_m, \ldots, c_m s_1 s_2 \ldots s_{m-1},\ s_1 s_2 \ldots s_m) = d = s_1 s_2 \ldots s_m.$$

$\therefore$ s_1 divides c_1, s_2 divides c_2, etc. so integral y exists and hence integral x.

7.1.3 $A^{-1}x + y = b \Leftrightarrow x + Ay = Ab$ which has all basic solutions integral for all b, so A^{-1} is totally unimodular by virtue of Theorem 7.5.

7.1.4 Possible 2×2 minors involving the two columns are

$$\begin{vmatrix} 1 & -1 \\ 1 & 1 \end{vmatrix} = 2, \quad \begin{vmatrix} 1 & 1 \\ 1 & -1 \end{vmatrix} = -2, \quad \begin{vmatrix} 1 & 1 \\ -1 & 1 \end{vmatrix} = 2,$$

$$\begin{vmatrix} -1 & 1 \\ 1 & 1 \end{vmatrix} = -2, \quad \begin{vmatrix} 1 & 1 \\ 1 & 1 \end{vmatrix} = 0, \quad \begin{vmatrix} 1 & -1 \\ -1 & 1 \end{vmatrix} = 0,$$

$$\begin{vmatrix} 1 & 1 \\ -1 & -1 \end{vmatrix} = 0, \quad \begin{vmatrix} -1 & -1 \\ 1 & 1 \end{vmatrix} = 0, \quad \begin{vmatrix} -1 & 1 \\ 1 & -1 \end{vmatrix} = 0,$$

$$\begin{vmatrix} 1 & 1 \\ 0 & 0 \end{vmatrix} = 0, \quad \begin{vmatrix} 1 & -1 \\ 0 & 0 \end{vmatrix} = 0, \text{ etc.}$$

First four are unacceptable, others imply stated conditions.

7.1.5 Can assume A nonsingular. All b_i are integers and unique solution is $x_1 = x_2 = \cdots = x_n = \frac{1}{2}$ so by Cramer's rule $\frac{1}{2} = \det D / \det A$, $\det D$ being an integer. Hence $\det A = 2 \det D$.

7.1.6 Let $d = \text{g.c.d.}\,(\Sigma\lambda_j a_{1j}, \Sigma\lambda_j a_{2j}, \ldots, \Sigma\lambda_j a_{nj})$ and put

$$b_i = \left(\sum_{j \in T} \lambda_j a_{ij}\right) / d.$$

The b_i are integers and unique solution of $Ax = b$ is $x_j = \lambda_j / d$, $j \in T$, $x_j = 0$, $j \notin T$. As in previous exercise, $\det A = (d \times \text{integer}) / \lambda_j$.

7.2.1

$$V(3, 3, 2) = \begin{bmatrix} U(3, 3) & 0 \\ 0 & U(3, 3) \\ I_9 & I_9 \end{bmatrix}$$

where

$$U(3,3) = \begin{bmatrix} 1 & 1 & 1 & 0 & 0 & 0 & 0 & 0 & 0 \\ 0 & 0 & 0 & 1 & 1 & 1 & 0 & 0 & 0 \\ 0 & 0 & 0 & 0 & 0 & 0 & 1 & 1 & 1 \\ & I_3 & & & I_3 & & & I_3 & \end{bmatrix}$$

and submatrix is

$$\begin{bmatrix} 1 & 1 & 0 & 0 & 0 & 0 & 0 \\ 1 & 0 & 1 & 0 & 0 & 0 & 0 \\ 0 & 0 & 0 & 1 & 1 & 0 & 0 \\ 0 & 0 & 0 & 0 & 0 & 1 & 1 \\ 0 & 0 & 0 & 1 & 0 & 0 & 1 \\ 0 & 1 & 0 & 0 & 1 & 0 & 0 \\ 0 & 0 & 1 & 0 & 0 & 1 & 0 \end{bmatrix}$$

which has determinant -2.

7.3.1 (See [10], p. 234.) If $B::A$, let $[I_m\ A]P = [G\ H]$ so from (7.3.8), $Q[G\ H] = [I\ B] \Rightarrow QG = I, B = QH = G^{-1}H$. Conversely, if G, H exist as stated $[I\ A]P = [G\ H] = [G\ GB]$ for some permutation matrix P. $\therefore G^{-1}[I\ A]P = [I\ B]$ so $B::A$.

7.3.2 (See [10], p. 235.) Let $C = -A^T$, $c_{ij} = -a_{ji}$. From (7.3.1),

$$C^{ji} = \frac{1}{c_{ji}} \begin{bmatrix} D_{11} & \cdots & -c_{1i} & \cdots & D_{im} \\ \cdot & \cdots & \cdot & \cdots & \cdot \\ \cdot & \cdots & \cdot & \cdots & \cdot \\ \cdot & \cdots & \cdot & \cdots & \cdot \\ c_{j1} & \cdots & 1 & \cdots & c_{jm} \\ \cdot & \cdots & \cdot & \cdots & \cdot \\ \cdot & \cdots & \cdot & \cdots & \cdot \\ \cdot & \cdots & \cdot & \cdots & \cdot \\ D_{n1} & \cdots & -c_{ni} & \cdots & D_{nm} \end{bmatrix}$$

where

$$D_{rs} = \begin{vmatrix} c_{rs} & c_{ri} \\ c_{js} & c_{ji} \end{vmatrix}, r \neq j,\ s \neq i$$

$$= \begin{vmatrix} -a_{sr} & -a_{ir} \\ -a_{sj} & -a_{ij} \end{vmatrix} = \begin{vmatrix} a_{sr} & a_{ir} \\ a_{sj} & a_{ij} \end{vmatrix} = d_{sr}$$

leading to $C^{ji} = -(A^{ij})^T$.

Subject Index

Author Index